T0300035

Medical Anthropology in Europe

This collection brings together three generations of medical anthropologists working at European universities to reflect on past, current and future directions of the field. Medical anthropology emerged on an international playing ground, and while other recently compiled anthologies emphasize North American developments, this volume highlights substantial ethnographic and theoretical studies undertaken in Europe. The first four chapters trace the beginnings of medical anthropology back into the two formative decades between the 1950s–1970s in Italy, German-speaking Europe, the Netherlands, France and the UK, supported by four brief vignettes on current developments. Three core themes that emerged within this field in Europe – the practice of care, the body politic and psycho-sensorial dimensions of healing – are first presented in synopsis and then separately discussed by three leading medical anthropologists, Susan Reynolds Whyte, Giovanni Pizza and René Devisch, complemented by the work of three early career researchers. The chapters aim to highlight how very diverse (and sometimes overlooked) European developments within this rapidly growing field have been, and continue to be. This book will spur reflection on medical anthropology's potential for future scholarship and practice, by students and established scholars alike.

This book was originally published as a special issue of *Anthropology & Medicine*.

Elisabeth Hsu is Professor in Anthropology at the University of Oxford, UK, where she co-founded and convenes the Medical Anthropology teaching and research programme.

Caroline Potter is Lecturer in Medical Anthropology at the University of Oxford, UK, where she obtained her doctorate in 2007 within this programme and assisted in building it up between 2008 and 2014.

Medical Anthropology in Europe

Shaping the Field

Edited by
Elisabeth Hsu and Caroline Potter

LONDON AND NEW YORK

First published 2015
by Routledge
2 Park Square, Milton Park, Abingdon, Oxon, OX14 4RN, UK

and by Routledge
711 Third Avenue, New York, NY 10017, USA

Routledge is an imprint of the Taylor & Francis Group, an informa business

British Library Cataloguing in Publication Data
A catalogue record for this book is available from the British Library

ISBN 13: 978-1-138-80800-3

Typeset in Times New Roman
by RefineCatch Limited, Bungay, Suffolk

Publisher's Note
The publisher accepts responsibility for any inconsistencies that may have
arisen during the conversion of this book from journal articles to book chapters,
namely the possible inclusion of journal terminology.

Disclaimer
Every effort has been made to contact copyright holders for their permission to
reprint material in this book. The publishers would be grateful to hear from any
copyright holder who is not here acknowledged and will undertake to rectify
any errors or omissions in future editions of this book.

Contents

CONTENTS

Citation Information

The chapters in this book were originally published in *Anthropology & Medicine*, volume 19, issue 1 (April 2012). When citing this material, please use the original page numbering for each article, as follows:

Please direct any queries you may have about the citations to
clsuk.permissions@cengage.com

Notes on Contributors

Serena Bindi, Université de Nice Sophia Antipolis, Nice, France; Centre for Himalayan Studies (Centre d'Etudes Himalayennes), Villejuif, France

René Devisch, Institute for Anthropological Research in Africa, KU Leuven – University of Leuven, Leuven, Belgium

Sjaak van der Geest, Medical Anthropology, University of Amsterdam, Amsterdam, The Netherlands

Elisabeth Hsu, Institute of Social and Cultural Anthropology, University of Oxford, Oxford, UK

Ruth Kutalek, Medical University of Vienna, Vienna, Austria

Gilbert Lewis, University of Cambridge, Cambridge, UK

Dominik Mattes, Institute of Social and Cultural Anthropology, Freie Universität Berlin, Germany

Verena C. Münzenmeier, Formerly University of Zurich, Zurich, Switzerland

Giovanni Pizza, Dipartimento Uomo & Territorio, University of Perugia, Italy

Caroline Potter, Institute of Social and Cultural Anthropology, University of Oxford, Oxford, UK

Armin Prinz, Medical University of Vienna, Vienna, Austria

Tullio Seppilli, Italian Society of Medical Anthropology (President) and Angelo Celli Foundation for a Culture of Health (President), University of Perugia, Italy

Koen Stroeken, African Cultures, Ghent University, Belgium

Susan Reynolds Whyte, Department of Anthropology, University of Copenhagen, Copenhagen, Denmark

Part I

Introduction to Part I

Medical anthropology in Europe: shaping the field

Elisabeth Hsu and Caroline Potter

Institute of Social and Cultural Anthropology, University of Oxford, Oxford, UK

'Europe' has been a point of reference for medical anthropologists previously (e.g. Lock 1986; DelVecchio Good et al. 1990; Pfleiderer and Bibeau 1991), and also in more recent years (e.g. Saillant and Genest 2007 [2005]).[1] In this issue, 'Europe' refers to teaching and research at European Universities, and their intellectual outreach. To be sure, a 'European' as opposed to a 'North American' or 'Japanese' medical anthropology does not exist. Nor are there distinctive national styles of doing medical anthropology; diversity prevails even within a single language community. Rather, medical anthropology has emerged as an academic field on an international playing ground, and trans-Atlantic exchanges have always drawn on a serious engagement with research in Asia, Africa, Meso- and South America. Considering that medical anthropology is now taught in a rapidly growing number of graduate and undergraduate courses in Europe (Hsu and Montag 2005), while recently compiled anthologies honour almost exclusively authors working in North America (e.g. Good, Fischer, and Willen 2010), this publication may provide a cautious 'corrective' (naturally with no claim to representativeness of all medical anthropology in Europe).

When one works within such a dynamic field that is also very polymorph, questions arise about how it all began. Reflections on origins are always linked to issues of self-definition and future developments. This special issue of *Anthropology & Medicine* attends to questions about the past and future in two parts. In the first part, six pioneers in the field (all currently in retirement) speak about their experiences when they started research and teaching, some 40 to 50 years ago: Tullio Seppilli, Gilbert Lewis, Jean Benoist, Sjaak van der Geest, Armin Prinz, and Verena Kücholl/Münzenmeier. Their essays capture pieces of oral histories of the times when there was not yet a field as such.

The second part of the special issue attends to current issues, and its introductory paper – asking *quo vadis?* – aims to identify core themes that have defined the field from the early days to the present. It identifies 'the practice of care' as one of the most well-researched themes by medical anthropologists in Europe, and explains this in light of the well-known tension in medicine between competence and care. Accordingly, studies into medical care, which often contained an implicit critique of medicalisation, countervailed earlier studies into 'the problem of knowledge' – a theme of research that has since radically changed, both in theoretical orientation and method. Early preoccupations with knowledge have now been superseded by research into practice and 'the body as a project in the making'.[2] The six papers on current issues are presented in three sections (further detailed in the introduction to Part II), whereby each presents the work of a senior scholar alongside a more junior one. The volume thus presents work from across three generations of medical anthropologists in Europe.

As Seppilli (this issue) demonstrates, an interest in a broader spectrum of themes relating to what is nowadays called medical anthropology can be traced back centuries and, in the twentieth century, as Van der Geest (this issue) shows, to tropical doctors' confrontation with their patients' health beliefs and practices in foreign countries. To these discussions others can be added, which highlight a distinctive point in time of striking convergences: the 1970s. The two essays on developments in Britain (Lewis this issue) and German-speaking countries (Kutalek, Münzenmeier, and Prinz this issue) demonstrate clearly that, during this formative period, the first interdisciplinary conferences were held on themes relevant to what was to emerge as a distinct field. The first monographs were then written, the first associations were founded (see vignette by Hardon and Beaudevin), the first journals were published (see Table 1), and the first university positions were established (both in anthropology departments and medical schools). In this respect, the history of the field in Europe was no different from that in North America.[3]

This phenomenon of the 1970s warrants an explanation.[4] Surely, post-war 'modern' medicine had reached a peak of its professionalisation then, and physicians had a status of respect and authority that they no longer have in the twenty-first century. Medical sociologists were articulate in their critique and fed into sociology's long-standing concern with 'modernity' and its trends towards normalisation and deviance, as initially discussed by Talcott Parsons and Erving Goffman. This sociological research, which was to have a profound impact on medical anthropology, was incorporated primarily – but not exclusively (e.g. Frankenberg 1980) – by way of the North American academy: Leslie's (1976) and Kleinman et al.'s (1975) edited volumes are both concerned with the professionalisation of medicine, but in South and East Asia respectively (rather than 'at home').

The focus on professionalisation brought with it an interest in 'health sectors', 'medical systems' and 'medical pluralism'. Of those, the first theme was to play a role particularly in public health research; the second instantly met with criticism, mostly by Africanists (e.g. Last 1981; Pool 1994); while the third, despite its bounded concept of culture and limited purview of power differentials in medical authority, has survived as a recurrent theme (e.g. the 2011 conference in Rome, organized by the EASA medical anthropology network).

However, the medical sociological orientations from North America met with another research interest then burgeoning at European universities: the study of social crises in medical terms, and their redress (see Lewis this issue). The description of society through the lens of crisis situations proved illuminating, as the workings of social institutions often go unnoticed otherwise; Turner's (1968) discussion of village and kinship politics carried out in a medical idiom was an acknowledged precursor. Accordingly, the 1960s and 1970s saw not only the medical profession's expansion and consolidation, critiqued by sociologists, but also the remarkable growth of and diversification in academic fields such as social/cultural anthropology, which

Table 1. Some well-known medical anthropology journals.

Journal	Dates of publication	Association
Etnoiatria: Rivista di Etnomedicina	1967–1968	Founding ed. Antonio Scarpa
Medical Anthropology Quarterly: International Journal for the Analysis of Health (formerly *Medical Anthropology Newsletter*)	1968–1982; 1983–1986; 1987–today	Organization of Medical Anthropology, est. 1967 (called Society for Medical Anthropology since 1970)
Ethnomedizin: Zeitschrift für interdisziplinäre Forschung	1971–1982	Arbeitsstelle für Ethnomedizin (AfE), est. 1969
Medical Anthropology: Cross-Cultural Studies in Health and Illness	1977–today	Founding eds. Robert Ness, Gretel H. Pelto and Pertti J. Pelto
Culture, Medicine & Psychiatry: An International Journal of Cross-Cultural Health Research	1977–today	Founding ed. Arthur Kleinman
Social Science and Medicine Part B: Medical Anthropology	1978–1981 (full journal dates: 1967–today)	Founding ed. Peter J.M. McEwan, Medical anthropology ed. Charles Leslie (1978–1989)
Curare: Ethnomedizin und Transkulturelle Psychiatrie	1978–today	Arbeitsgemeinschaft Ethnomedizin (AGEM), est. 1970
Bulletin d'Ethnomédicine (formerly *Bulletin de Liason, Seminaire Mensuel d'Anthropologie Médicale*)	1980/82–1988	Founding ed. A. Epelboin
Anthropology & Medicine (formerly *Newsletter of the BMAS*)	1981–1993; 1994–today	British Medical Anthropology Society (BMAS), est. 1976
Antropologia Medica: Per un Confronto di Culture sui Temi della Salute	1986–1988	Gruppo di antropologia della salute e della malattia, Istituto di Igiene della Università di Trieste
Bulletin Amades	1988–today	Anthropologie Médicale Appliquée au Développement et à la Santé (AMADES), est. 1988
Medische Antropologie: Journal about Culture and Health	1989–today	Medical Anthropology Group, University of Amsterdam, in collaboration with Katholieke Universiteit Leuven
AM Rivista della Societa Italiana di Antropologia Medica	1996–today	Societa Italiana di Antropologia Medica (SIAM), est. 1988
Viennese Ethnomedicine Newsletter	1998–today	Unit Ethnomedicine and International Health, Medical University of Vienna, est. 1993

responded to a rapidly changing world. The ethnographic documentation of acute crisis situations, and the social processes they instigated, appear to have laid the foundations for the field that we nowadays call medical anthropology.

Medical anthropologists have regularly engaged with research in public health and primary care, both abroad and at home. It is probably an achievement, even if sometimes feelings to the contrary arise, that the field has been able to accommodate a most heterogeneous group of researchers. Some Social Anthropology departments established tenured positions in medical anthropology, such as those in Cambridge, Amsterdam, Leuven and Copenhagen. In other places, the School of Medicine installed short-term training programmes for public health professionals intent on doing development work 'in the tropics', such as those in Heidelberg and Aix-en-Provence. In the late 1980s, the first master's courses were set up, catering primarily to general practitioners interested in the socio-cultural aspects of the medical encounter, such as those in Oslo and Brunel. And finally, there are those centres that have done all three for over half a century, such as that in Perugia.

In Europe, as presumably also elsewhere, so-called 'applied' medical anthropologists are the most numerous, and for many medical anthropologists, this remit has been the field's main strength (Helman 2006 – his introductory book (Helman 2007 [1984]) is now in its fifth edition and has been translated into several languages). Applied medical anthropologists' comparatively high employment rates have come at a price, however. In some settings, Rapid Assessment Procedures (RAPs) and Participatory Rural Appraisals (PRA) have replaced language-competent, long-term fieldwork – deemed too unspecific, too time-consuming and too costly to be implemented in the light of the 'urgent needs' that medical practitioners alleviate.

Two themes, directly relevant to anthropology at home, have more recently marked the 'applied' field. Neither is extensively discussed in this volume, but they are briefly mentioned here by way of outlining the horizons within which this volume is situated. The first centres on changing notions of 'citizenship', constructed in relation to one's biomedically defined and hence 'biological' status (Petryna 2002). This emergent research relates the biomedical to the legal and bureaucratic, often framing issues of morality in terms of human rights. Medical questions come to serve as vehicles for the wider political commentary (e.g., which lives to save during war; Fassin 2007). The second 'applied' theme concerns research into complementary and alternative medicine (CAM), otherwise mostly undertaken by health professionals in primary care. Not least in response to specifically post-socialist developments, Eastern European researchers have become particularly active in studying CAM, also as expatriates (e.g. Penkala-Gawecka 2002; Johannessen and Lazar 2006; Lindquist 2006). Their research converges in interesting ways with CAM research undertaken in Mediterranean countries and Portugal, which appears to have arisen out of a longstanding association with folklore studies, as evidenced in a recent volume on new age versions of age-old bathing cultures (Naraindras and Bastos 2011).

As the contributors to this special issue demonstrate, however, the field has always engaged in substantial theoretical developments, in addition to its value for people working in 'applied' structures. At this juncture of medical anthropology's recent and ever-wider expansion in Europe, the authors call for a moment of reflection, so that we might position our current research endeavours in respect of our own past.

Acknowledgements

This special issue is based upon papers presented at the conference 'Medical Anthropology in Europe' funded by the Wellcome Trust and Royal Anthropological Institute.

Conflict of interest: none.

Notes

1. See also Diasio (1999) on central European developments and Ingstad and Talle (2009) on the Nordic network of medical anthropology.
2. Admittedly, the mind (and mental illness) rather than the body figured among the early core themes, but since the body is central to the medical anthropology programme at Oxford, where the RAI conference that prompted this special issue took place, it provided a thematic anchor.
3. European appointments in medical anthropology included those at Perugia (Seppilli, from 1956), Cambridge (Lewis, from 1971), Heidelberg and Zurich (assistant lecturers, from 1977), Amsterdam (Van der Geest, from 1978), and Aix-en-Provence (Benoist, from 1981). These initiatives broadly coincided with North American appointments at McGill (Lock), Harvard (Kleinman, Good) and UC-Berkeley (Leslie, Scheper-Hughes).
4. Beginnings in the 1970s appear to apply only to countries where social anthropology was already an established discipline of higher education. In socialist Croatia, biological anthropologists founded an anthropological society in 1977 and a 'Commission on Medical Anthropology and Epidemiology' in 1988, while possibilities for developing a critical medical anthropological perspective arose only after the decline of socialism (Spoliar-Vrzina, in Hsu and Montag 2005: 7). In Spain, Ackerknecht's 'culturalism', as adopted by medical historians and folklorists, predominated in discussions when medical anthropology was first instituted in a taught course (in 1981), well before social/cultural anthropology became firmly established (in the early 1990s according to Comelles, Perdiguero, and Martínez-Hernáez (2007 [2005]: 108)). By contrast, in Zurich, when *Ethnologie* rather than *Völkerkunde* was being instituted, Ackerknecht was not once mentioned in the first medical anthropology lectures, although he worked until his retirement in 1971 *im Turm* – 'in the lofty tower' of the University's main building, only two flights of stairs above the classroom where the lectures were held (a co-author's observation).

References

Comelles, Josep M., Enrique Perdiguero, and Angel Martínez-Hernáez. 2007 [2005]. Topographies, folklore, and medical anthropology in Spain. In *Medical anthropology: Regional perspectives and shared concerns*, eds. Francine Saillant and Serge Genest, 103–21. Malden: Blackwell.

DelVecchio Good, M.J., D.R. Gordon, M. Pandolfi, and B.J. Good. 1990. Traversing boundaries: European and North American perspectives on medical and psychiatric anthropology. *Culture, Medicine and Psychiatry* 14, no. 2: 141–4.

Diasio, Nicoletta. 1999. *La science impure: Anthropologie et médecine en France, Grande-Bretagne, Italie, Pays-Bas*. Paris: Presses Universitaires de France.

Fassin, D. 2007. Humanitarianism as a politics of life. *Public Culture. Society for Transnational Cultural Studies* 19, no. 3: 499–520.

Frankenberg, R. 1980. Medical anthropology and development: A theoretical perspective. *Social Science and Medicine* 14B: 197–207.

Good, Byron, Michael Fischer, and Sarah Willen, eds. 2010. *A reader in medical anthropology: Theoretical trajectories, emergent realities*. Chichester: Wiley-Blackwell.

Helman, C. 2006. Why medical anthropology matters. *Anthropology Today* 22, no. 1: 3–4.

Helman, Cecil G. 2007 [1984]. *Culture, health and illness*. 5th ed. London: Hodder Arnold.

Hsu, Elisabeth, and Doreen Montag, eds. 2005. *Medical anthropology in Europe: Teaching and doctoral research*. Wantage: Sean Kingston.

Ingstad, Benedicte, and Aud Talle, eds. 2009. *Introduction to Nordic anthropology*. Special Issue, *Medical Anthropology Quarterly* 23, no. 1: 1–87.

Johannessen, Helle, and Imre Lazar, eds. 2006. *Multiple medical realities: Patients and healers in biomedical, alternative and traditional medicine*. Oxford: Berghahn Books.

Kleinman, A., P. Kunstadter, E.R. Alexander, and J.L. Gale, eds. 1975. *Medicine in Chinese cultures: Comparative perspectives*. Washington, DC: USGPO for Fogarty International Center, NIH.

Last, M. 1981. The importance of knowing about not knowing. *Social Science and Medicine* 15B, no. 3: 387–92.

Leslie, Charles, ed. 1976. *Asian medical systems*. Berkeley: University of California Press.

Lindquist, Galina. 2006. *Conjuring hope: Healing and magic in contemporary Russia*. Oxford: Berghahn Books.

Lock M., ed. 1986. Sympósion: Medical anthropology in Europe: The state of the art. *Medical Anthropology Quarterly* 17, no. 4: 87–95.

Naraindras, Harish, and Cristiana Bastos, eds. 2011. *Healing holidays? Itinerant patients, therapeutic locales and the quest for health*. Special Issue, *Anthropology & Medicine* 18, no. 1: 1–144.

Penkala-Gawecka, D. 2002. Korean medicine in Kazakhstan: Ideas, practices and patients. *Anthropology & Medicine* 9, no. 3: 315–36.

Petryna, Adriana. 2002. *Life exposed: Biological citizens after Chernobyl*. Princeton: Princeton University Press.

Pfleiderer, Beatrix, and Gilles Bibeau, eds. 1991. *Anthropologies of medicine: A colloquium on West European and North American perspectives*, *Curare* Special Volume 7. Heidelberg: Vieweg.

Pool, Robert. 1994. *Dialogue and the interpretation of illness: Conversations in a Cameroon village*. Oxford: Berg.

Saillant, Francine, and Serge Genest, eds. 2007 [2005]. *Medical anthropology: Regional perspectives and shared concerns*. Malden: Blackwell.

Turner, Victor. 1968. *The drums of affliction: A study of religious processes among the Ndembu of Zambia*. Oxford: Clarendon Press.

Vignette

Medical anthropologists in Europe connect

There are a variety of regional and thematic medical anthropology associations in Europe (see vignettes and Note 1), some of which publish a journal (see Table 1). The Medical Anthropology Network is a recent association, established in 2006 and run by an elected committee. It operates within the European Association of Social Anthropologists (EASA) and specifically aims to provide information on seminars, research collaborations, publications and graduate education (through its list-serve MedAnthNet), and to create opportunities for medical anthropologists and graduate students in Europe to connect with each other. The Medical Anthropology Network members meet at the biennial EASA conferences and at other events hosted by the various regional sub-groups. In September 2011 it held its first independent meeting in Rome around the theme 'Medical pluralism: Techniques, politics, institutions'. The Network also includes the Medical Anthropology Young Scholars (MAYS), which organises annual seminars and connects 300 young scholars from European universities and research institutions. In 2013, the European Medical Anthropology

Network will co-organise the Second International Independent Medical Anthropology Conference together with the American Society for Medical Anthropology (SMA).

Anita Hardon
Current chair of the network
Claire Beaudevin
Network committee member in charge of MAYS

http://dx.doi.org/10.1080/13648470.2012.688344

Alien origins: xenophilia and the rise of medical anthropology in the Netherlands

Sjaak van der Geest

Medical Anthropology, University of Amsterdam, Amsterdam, The Netherlands

The beginnings of medical anthropology in the Netherlands have a 'xenophile' character in two respects. First, those who started to call themselves medical anthropologists in the 1970s and 1980s were influenced and inspired not so much by anthropological colleagues, but by medical doctors working in tropical countries who had shown an interest in the role of culture during their medical work. Secondly, what was seen as medical anthropology in those early days almost always took place in 'foreign' countries and cultures. One can hardly overestimate the exoticist character of medical anthropology up to the 1980s. It was almost automatic for anthropologists to take an interest in medical issues occurring in another cultural setting, while overlooking the same issues at home. Medical anthropology 'at home' started only around 1990. At present, medical anthropology in the Netherlands is gradually overcoming its xenophile predilection.

The first Dutch study that explicitly referred to 'medical anthropology' appeared in 1964.[1] It was a dissertation by a medical doctor, Vincent van Amelsvoort, on the introduction of 'Western' health care in the former Dutch colony of New Guinea (now an uneasy province of Indonesia). It came only one year after Norman Scotch had delineated medical anthropology as a formal field of research and teaching, and as a sub-discipline of cultural anthropology.[2] Van Amelsvoort's (1964a) study focused on the clash between two entirely different (medical) cultures, and it was quickly followed by the publication of a short note on the new field of medical anthropology in a Dutch medical journal (Van Amelsvoort 1964b). Discussing the origins of the new sub-discipline, Van Amelsvoort referred mainly to social scientists and health professionals who (like himself) had worked in the field of health development and had analysed the relationship between culture, health and health practices: Charles Erasmus, Edward Wellin, Walsh McDermott and G. Morris Carstairs. Van Amelsvoort was a 'tropical doctor' with a keen interest in culture, thrust upon him during his work as a colonial doctor in New Guinea. Later he became professor in 'Health Care in Developing Countries' in the medical faculty of the University of Nijmegen. The biographical background of his work in medical anthropology typifies the 'alien' origins of the field in the Netherlands, referring to both disciplinary and geographical territories. Medical anthropology in the

Netherlands started in tropical areas far from home, and those who initiated it were not anthropologists but medical doctors.

Medical initiatives

There are at least three reasons why physicians and not anthropologists took up the issue of culture and medicine. First, the social and cultural character of health problems manifests itself much more prominently in medical practice than in anthropological research. In their attempts to improve health conditions, tropical doctors continuously encountered 'cultural barriers'. It forced them to think about the nature of these barriers and to reflect on their own mission. Whatever opinion they developed about the practical implications of those cultural barriers, many of them at least realised that it was crucial to learn more about them. There was a need for knowledge about local cultures, particularly medical cultures.

That awakening of cultural interest among Dutch tropical doctors can be observed in the work of some early physicians in the Dutch Indies (later Indonesia). J.P. Kleiweg de Zwaan (1910) published a study about indigenous medicine among the Menangkabau people in North Sumatra, J.M. Elshout (1923) did the same about the Dayak people in Borneo, and J.A. Verdoorn (1941) wrote a study about indigenous midwifery in various ethnic groups of the colony. Another colonial precursor of medical anthropology was F.D.E. van Ossenbruggen, a lawyer who was particularly interested in how illness and health practices were embedded in the general culture of local Indonesian groups. His work includes a comparative study of rituals against smallpox among different populations (Van Ossenbruggen 1916; see also Diasio 2003, Niehof 2003).

Van Amelsvoort's dissertation in 1964 followed the tradition of colonial doctors, as did the study by Gerard Jansen (1973) on doctor-patient relationships in Bomvanaland (South Africa). Jansen had spent 11 years as a missionary doctor in Bomvana society. It was around that time, in the 1970s, that Dutch anthropologists became interested in the cultural identity of health and medicine and 'took over' the job of medical anthropology from their medical colleagues.

The applied purpose of medical anthropology remained strong, however, after anthropologists became involved in the work. Many of the first *anthropological* medical anthropologists in the Netherlands worked in close cooperation with – or in the service of – medical projects. The anthropologist Douwe Jongmans, for example, moved from the University of Amsterdam to the health section of the Royal Tropical Institute and did research among North African immigrants. His focus was on cultural perceptions and practices around fertility and birth regulation (Jongmans 1974, 1977). Several other anthropologists continued (in varying degrees) to practice 'anthropology *in* medicine' – mostly abroad, but increasingly also in the Netherlands, albeit mainly among migrants.

A second explanation for the 'failure' of anthropologists to grasp the opportunity of medical anthropology at an earlier stage could be their weariness of so-called 'applied anthropology', which dominated the post-colonial era of anthropology. In the 1950s and 1960s, most anthropologists fostered – as far as possible – the principle of non-intervention. 'Proper' anthropologists, it was believed, should not make their hands dirty on government- or mission-initiated development projects. Problems of illness and death were to be studied primarily as occasions for social conflict or religious ceremony. Individual cases of illness and death *per se* did not really interest them. Only when they occurred in their immediate environments and touched them personally were they likely to become more actively involved. Many anthropologists, for example, distributed medicines to members of 'their family' and to neighbours, and helped them in other ways. Some anthropologists were known to 'play doctor' and even held 'consulting hours'. Such activities generally remained separate from their ethnographic work, however, and did not lead them to anthropological reflection. Not only did these activities fall outside the scope of their research, but they were also in conflict with the 'rules' of proper anthropological fieldwork: non-intervention, participant *observation*, with the emphasis on the second word.

Evans-Pritchard is a well-known example of an anthropologist busily engaged in medical activities, but he was also an exception. In his Azande study, Evans-Pritchard (1937, 506) writes that he was 'constantly associated with every kind of sickness.' When his medical work became more cumbersome and took him two hours every morning, he arranged for assistance; someone else came to dress wounds and dispense medicines for him. It had not been a waste of time. He writes:

> When...I generalize about Zande notions of disease, I do so on a fairly wide experience. I have invariably found that when a Zande is struck down by general and acute sickness, with sudden and severe symptoms and rapid course, as in certain types of fever, pneumonia, cerebrospinal meningitis, influenza, &c, his relatives and neighbours straightaway connect his collapse with the primary cause of witchcraft or sorcery...(Evans-Pritchard 1937, 506)

His reflection on the mundane work of treating sick people enabled him – as an anthropologist – to write the first major work in medical anthropology, decades before medical anthropology emerged as an academic discipline.

Finally, it must be remembered that the birth of anthropology around the turn of the nineteenth into the twentieth century was partly a reaction to the growing hegemony of biology and its excrescences into flat evolutionism, racism and eugenics. This 'classic' anthropological 'allergy' to biology probably added to anthropologists' initial reluctance to get involved in medical issues. It was only in the 1970s that anthropologists 'discovered' the symptoms of bodily dysfunction as cultural phenomena and became fascinated by medical topics. That was the moment when medical anthropology in the Netherlands – as in many other countries – became a recognised and popular field of study within cultural and social anthropology.[3]

Xenophilia

The other type of alienism during the first years of medical anthropology in The Netherlands is geographical. Research that was recognised as 'medical anthropology' had always taken place far away, on foreign territory. This is not surprising, because at that time anthropology was seen as the study of 'other cultures' (Beattie 1964). With some exaggeration one could say that not the topic but the topos made a study 'anthropological'. Studies on social and cultural aspects of health, body, mind, emotion and well-being, which were situated in Dutch society were not considered 'anthropological' *because* they dealt with Dutch issues; they were not even noticed by anthropologists. Conversely, work done under the tropical sun was embraced as anthropology or anthropologically relevant, even if it was rather far removed from anthropology in theoretical and methodological respects.

Two examples illustrate this point.[4] The pioneering studies of the physician, biologist, psychologist and philosopher F.J.J. Buytendijk (1887–1974) is hardly ever referred to in the publications of the early Dutch medical anthropologists. Buytendijk's main concern can be characterised as a consistent attempt to overcome the body/mind dichotomy, a theme which, about 30 years later, became the inspiration for one of the most influential publications in medical anthropology (Scheper-Hughes and Lock 1987). Using data from physiology and ethology, Buytendijk tried to make the ideas of the French philosopher Merleau-Ponty about the 'body-subject' plausible and acceptable to a forum of hard scientists. He argued for an 'anthropological physiology': a physiology that – as Merleau-Ponty suggested – was shown to react meaningfully to human experience. He applied his views to bodily reactions such as sleeping and being awake, pain, being thirsty, blushing, sweating and fainting. Buytendijk felt closely affiliated to the Heidelberg group in Germany where Viktor von Weizsäcker, Herbert Plügge, Thure van Uexküll and others attempted to develop a non-dualistic brand of medicine.[5] Buytendijk, whose work has been translated into English, shows that there is subjectivity and meaningful 'behaviour' in physiological processes. The body is a cultural actor, and bodily dysfunction is a meaningful cultural act (Buytendijk 1974). As stated above, Buytendijk's publications were not thought to be relevant to cultural anthropologists. In fact, that negligence was mutual. Buytendijk took his inspiration and data from biology and psychology, about human beings as well as animals, but never referred to studies of people in other cultures. It is doubtful that he read anthropological work. This may look an amazing omission with hindsight, but we should realise that at that time anthropologists offered little on embodied processes that could have interested Buytendijk.[6]

A similar story can be told about the Dutch psychiatrist J.H. van den Berg. Outside the Netherlands, Van den Berg is best known for a brief treatise on the psychology of the sick-bed (Van den Berg 1966 [1954]), which has been translated into many languages. His phenomenological description of the experience of the bed for a sick person is surprisingly anthropological. The bed is a safe haven for a healthy person, an intimate place where he can rest, recharge his energy, be alone or make love. It is a place full of promise. But for the sick person the bed may become a prison, the place that he wants to leave but cannot. For the chronically ill person in particular, the bed becomes the symbol of a life without future. The sounds that reach him from the street remind him of the world that he lost. This beautiful emic capture of the illness experience was eagerly read by generations of nurses throughout the 'Western' world, but remained unnoticed by anthropologists for a long time.

Within his own country, Van den Berg drew considerable attention through his book *Metabletica* (1956), a study of societal changes in a historical perspective. Some years later he published his monumental study of the human body from a 'metabletic' perspective (Van den Berg 1959, 1961).[7] His main thesis was that the human body has changed through the ages (his study takes the reader back to the thirteenth century). He not only argues that the *meaning* of the body varies over time, but also that the body itself, 'in its materiality', has changed. Van den Berg's style of reasoning does not fit in any conventional discipline, and one could characterise him perhaps as a 'postmodernist avant la lettre'. His argument follows unpredictable associations, from paintings by Brueghel, Rubens and Picasso, to a mystic's vision, a book of devotion, a scientific study of the heart, a paper clipping about the rescue of a drowning person, a collection of lyrics, an X-ray photograph, and a building by Le Corbusier. The body, Van den Berg writes, reflects the ideas and politics of its period. Again, this is a viewpoint busily discussed by anthropologists today, but it went unnoticed by them at the time. Conversely, again, it should be noted that Van den Berg showed no interest in anthropologists' descriptions of human bodies in other cultures. The xenophilia of the anthropologists paralleled the 'xenophobia' of other disciplines that occupied themselves with body, culture and society.

One could perhaps say that at present Dutch medical anthropology is alien-oriented in yet a third way. Literature read in its teaching courses is overwhelmingly foreign, demonstrating an extreme form of non-chauvinism. Dutch authors are hardly mentioned in the most popular handbooks and readers of medical anthropology. The most ambitious study on the foundations of medical anthropology written by a Dutch author is entirely devoted to a debate with the American school of Kleinman and hardly touches on the achievements of the 'Dutch school' (Richters 1991).

In sum, the origins of Dutch medical anthropology are in two respects 'alien', in a disciplinary and in a geographical sense. The latter is the dominant alienation. One cannot overestimate the exoticist character of anthropology – and medical anthropology in particular – up to the 1980s. It was almost automatic for anthropologists to take an interest in anything happening in another culture and to overlook anything happening at home. This predilection for 'things from far', exoticism in brief, was of course an inverted type of ethnocentrism: 'culture', the object of anthropological scrutiny, was only to be found among the 'others', while at home they had science, medicine and the Christian faith, untainted by the relativist adjective 'cultural' (see further: Van der Geest 2002). The only things occurring in another culture that did not interest them were events or institutions that reminded them of home and did not fit their conception of 'culture'. Schools, Christian churches, western-type hospitals and health centres were skipped or blotted out of their ethnographic work; they did not observe them nor participate in them (Van der Geest and Kirby 1992).

If Buytendijk had written his treatise on phenomenological physiology in Borneo, anthropologists would have embraced him as a colleague. If Van den Berg had written about the sickbed of patients in Congo, the same would have happened.

Medical anthropology in the Netherlands today, following mainstream cultural anthropology, tries to overcome its 'alien' beginnings and come 'home' (on Medical Anthropology At Home, see the vignette by Sylvie Fainzang). This trend is probably stimulated also by the changing epidemiological scene in its society. Chronic disease and old age take an increasing amount of attention. The emphasis shifts from active medical intervention to care and social attention. The present popularity of medical anthropology among students has been surprising. The author's impression is that medical anthropology caters for two types of students. It continues to be a branch of anthropology that 'matters' in the sense that it can be applied in practical medical and paramedical work. But medical anthropology also has become a major domain of theorising about culture. Health and illness, body and food, care and violence, anatomy and genetics, medical science and medical hegemony constitute excellent cases to explore the 'work of culture'. This 'double identity' (applied and theorising) typifies medical anthropology as it is presently practiced in the Netherlands and at the University of Amsterdam in particular.

Acknowledgements

The author thanks his colleagues at the University of Amsterdam, Elisabeth Hsu and Gilbert Lewis for their suggestions. The paper, which draws on an earlier publication on this theme (Van der Geest 2006), was presented at the conference 'Medical Anthropology in Europe' funded by the Wellcome Trust and Royal Anthropological Institute.

Conflict of interest: none

Notes

1. The most elaborate discussion of past and present Dutch medical anthropology is, perhaps not surprisingly, from an outsider: the Italian anthropologist Diasio (1999, 2003), who studied medical anthropological traditions in four European societies (France, Great Britain, Italy and The Netherlands). She writes that Dutch medical anthropologists see themselves as a mixed breed of foreign and interdisciplinary influences. This paper supports her thesis on this mixed provenance.
2. A birth date of medical anthropology does not exist, but 1953 was undoubtedly an important year for the contribution by Caudill (a psychiatrist by training) to Kroeber's *Anthropology today* about 'Applied anthropology in medicine' (Caudill 1953). Ten years later, Scotch published his overview of medical anthropological work, which began with the premise that '...in every culture there is built around the major life experiences of health and illness a substantial and integral body of beliefs, knowledge and practices' (Scotch 1963, 30). It was one of the first attempts to define the object of study of medical anthropology.
3. Medical anthropology developed along similar lines in other countries. The most prominent ancestors of medical anthropology in Britain, for example, were physicians (Rivers, Lewis, Loudon) and the same goes for the USA (Ackerknecht, Paul, Kleinman). For the medical roots of British medical anthropology, see Diasio (1999, 122–44).
4. Another much earlier example is the work of Dutch hygienists in the nineteenth century, in particular that of Pruys van der Hoeven, who emphasised the social and political nature of health and disease. Richters (1983) and Diasio (1999, 2003) discuss the hygienists' (unrecognised) link with medical anthropology.
5. In Heidelberg, the term 'medical anthropology' (*Medizinische Anthropologie*) was used long before the word was introduced in the Anglophone world, but it had another meaning: the philosophical reflection on illness, health and healing (cf. Von Weizsäcker 1927). As a consequence, German medical anthropologists were unable to adopt the term, as it already had another destination. They are still struggling for a decent name to the discipline which their colleagues outside Germany term 'medical anthropology'.
6. A good overview of work produced by the Dutch phenomenological school is Kockelmans (1987).
7. See also Van den Berg (1987).

References

Beattie, J. 1964. *Other cultures. Aims, methods and achievements in social anthropology*. London: Routledge.

Buytendijk, F.J.J. 1974. *Prolegomena to an anthropological physiology*. Pittsburgh: Duquesne University Press (original Dutch version: 1965).

Caudill, W. 1953. Applied anthropology in medicine. In *Anthropology today*, ed. A.L. Kroeber, 771–806. Chicago: University of Chicago Press.

Diasio, N. 1999. *La science impure: Anthropologie et médecine en France, Grande-Bretagne, Italie, Pays-Bas*. Paris: Presses Universitaires de France.

Diasio, N. 2003. 'Traders, missionaries and nurses', and much more: Early trajectories towards medical anthropology in The Netherlands. *Medische Antropologie* 15, no. 2: 263–86.

Elshout, J.M. 1923. *Over de geneeskunde der Kenja–Dajak in Centraal–Borneo in verband met hun godsdienst* [About the medical tradition of the Kenja-Dajak in Central Borneo in relation to their religion]. Amsterdam: Johannes Müller.

Evans-Pritchard, E.E. 1937. *Witchcraft, oracles and magic among the Azande*. Oxford: Clarendon Press.

Jansen, G. 1973. *The doctor–patient relationship in an African tribal society*. Assen: Van Gorcum.

Jongmans, D.G. 1974. Socio-cultural aspects of family planning: An anthropological study at the village level. In *The neglected factor. Family planning: Perception and reaction at the base*, eds. D.G. Jongmans and H.J.M. Claessen, 33–64. Assen: Van Gorcum.

Jongmans, D.G. 1977. Gastarbeider en gezondheidszorg. Eer en zelfrespect: De Noord-Afrikaanse boer en de overheid [Migrant worker and health care. Honour and self-respect: The North-African peasant and the authorities]. *Medisch Contact* 32: 509–12.

Kleiweg de Zwaan, J.P. 1910. *De geneeskunde der Menangkabau–Maleiers* [The medical tradition of the Menangkabou Malay]. Amsterdam: Meulenhoff.

Kockelmans, J.J., ed. 1987. *Phenomenological psychology: The Dutch school*. Dordrecht: Martinus Nijhoff Publications.

Niehof, A. 2003. The Indonesian archipelago as nursery for Leiden anthropology: Supplementary notes to Nicoletta Diasio. *Medische Antropologie* 15, no. 2: 292–95.

Richters, A. (J.M.) 1983. De medische antropologie: Een nieuwe discipline? [The medical anthropology: A new descipline?] *Antropologische Verkenningen* 2, no. 3: 39–69.

Richters, A. (J.M.) 1991. *De medisch antropoloog als verteller en vertaler. Met Hermes op reis in het land van de afgoden* [The medical anthropologist as narrator and translator: Travelling with Hermes in the land of idols]. Delft: Eburon.

Scheper-Hughes, N., and M.M. Lock. 1987. The mindful body: A prolegomenon to future work in medical anthropology. *Medical Anthropology Quarterly NS* 1, no. 1: 6–41.

Scotch, N.A. 1963. Medical anthropology. *Biennial Review of Anthropology* 1963: 30–68.

Van Amelsvoort, V.F.P.M. 1964a. *Early introduction of integrated rural health into a primitive society. A New Guinea case study in medical anthropology*. Assen: Van Gorcum.

Van Amelsvoort, V.F.P.M. 1964b. Medische antropologie, een terreinverkenning [Medical anthropology, an exploration]. *Nederlands Tijdschrift voor Geneeskunde* 108: 1289–90.

Van den Berg, J.H. 1956. *Metabletica of leer der veranderingen* [Metabletica or the theory of changes]. Nijkerk: Callenbach.

Van den Berg, J.H. 1959. *Het menselijk lichaam. Een metabletisch onderzoek* [The human body: A metabletic investigation]. Volume 1: *Het geopende lichaam* [The open body]. Nijkerk: Callenbach.

Van den Berg, J.H. 1961. *Het menselijk lichaam. Een metabletisch onderzoek* [The human body: A metabletic investigation]. Volume 2: *Het verlaten lichaam* [The deserted body]. Nijkerk: Callenbach.

Van den Berg, J.H. 1966. *The psychology of the sickbed*. Pittsburgh: Duquesne University Press [Dutch original 1954].

Van den Berg, J.H. 1987. The human body and the significance of human movement: A phenomenological study. In *Penomenological psychology: The Dutch School*, ed. J.J. Kockelmans, 55–78. Dordrecht: Nijhoff.

Van der Geest, S. 2002. Introduction: Ethnocentrism and medical anthropology. In *Ethnocentrism: Reflections on medical anthropology*, eds. S. van der Geest and R. Reis, 1–23. Amsterdam: Aksant.

Van der Geest, S. 2006. A cultural fascination with medicine: Medical anthropology in The Netherlands. In *Medical anthropology: Regional perspectives and shared concerns*, eds. F. Saillant and S. Genest, 162–82. Oxford: Blackwell.

Van der Geest, S., and J.P. Kirby. 1992. The absence of the missionary in African ethnography, 1940–1965. *African Studies Review* 35, no. 3: 59–103.

Van Ossenbruggen, F.D.E. 1916. Het primitieve denken, zoals dat zich uit voornamelijk in pokkengebruiken op Java en elders. Bijdrage tot de pre-animistische theorie [Pimitive thought as expressed in customs around smallpox in Java and elsewhere: A contribution to pre-animistic theory]. *Bijdragen tot de Taal, -Land- en Volkenkunde van Nederlandsch-Indië* 71: 1–370.

Verdoorn, J.A. 1941. *Verloskundige hulp voor de inheemsche bevolking van Nederlandsch–Indië: Een sociaal medische studie* [Obstetric help for the indigenous population of Dutch Indies: A socio-medical study]. 's–Gravenhage: Boekencentrum.

Von Weizsäcker, V. 1927. Ueber medizinische Anthropologie [About medical anthropology]. *Philosophischer Anzeiger* 2: 236.

Vignette

Medical anthropology at home

MAAH is an international network of medical anthropologists who do research in their own societies. It was born in 1998, after the first European Conference on 'Medical Anthropology at Home' held in the Netherlands, at the initiative of the University of Amsterdam. It stemmed from the belief that medical anthropology had for too long neglected the study of our own societies, towards which anthropologists had turned only recently, and that it would be fruitful to promote a collective thinking about the theoretical and methodological implications of research in this context. Its broader aim is to bring together medical anthropologists in order to discuss theoretical, methodological and practical issues in relation to health and culture, and to reinforce the position of medical anthropology in Europe. Although most of its members belong to European societies, it also welcomes researchers from other continents, with about 15 countries represented. The network organises regular conferences in various countries, during which research is discussed on themes such as the body, reproduction, drug use, doctor/patient relationships, chronic illness, health systems, medical pluralism, multiculturalism, migrations, and political engagement. But MAAH events also allow for discussions of methodological and epistemological issues connected to the choice of working in familiar settings, and they have encouraged reflection on the relationships between anthropology and medicine. These conferences give rise to collective publications of books and special issues of journals (including *Anthropology & Medicine*, *Antropologia Medica*, Aarhus Press, URV Publications), which echo these exchanges. The network has an international scientific Advisory Committee and a mailing list [see http://www.vjf.cnrs.fr/maah-france/].

Sylvie Fainzang

http://dx.doi.org/10.1080/13648470.2012.688347

Itineraries and specificities of Italian medical anthropology

Tullio Seppilli

Italian Society of Medical Anthropology (President) and Angelo Celli Foundation for a culture of health (President), University of Perugia, Italy

This paper describes the birth (or rebirth) of Italian medical anthropology around the middle of the 1950s, and its subsequent complex development up to the present. During this fairly long process, the author played a role that was probably of some importance, that of both a direct witness and active participant. Here these developments are briefly reviewed, in an attempt to single out some of the stimuli and the most significant occasions that have happened, their theoretical and methodological reference points, the main lines of research that have been tackled along the way, as well as the 'social demand' and the 'social use' that have integrated and oriented the practice of the new discipline within the horizon of some of the more general problems of Italian society. In outlining here the profile of and the various events in Italian medical anthropology, this paper takes into account the fact that, although a medical anthropology with that name and the disciplinary set-up that are now internationally attributed to it began in Italy only in the mid-1950s, important lines of research to which we would today attach that name had been undertaken long ago.

1. Exercises in medical anthropology before the advent of medical anthropology

It is not possible in a brief contribution to take account of what we might call the 'precedents' of medical anthropology. But it is perhaps useful here to outline at least the essential junctures in order, if for no other reason, to clarify what we are dealing with. Taking everything into consideration, in a period ranging between the first half of the seventeenth century and the first half of the twentieth, it seems possible to single out five main, firmly-rooted 'moments', albeit with different weights and roles in their respective socio-historical contexts. This paper will briefly trace out their main features.

1.1. *The first half of the seventeenth century: the 'reports' to the Florentine Health Magistracy, i.e. to the central health authority of the Grand Duchy of Tuscany*

These 'reports' were the fruit of surveys carried out by qualified medical practitioners that the Magistracy sent out to the various territories of the Duchy, when it considered that there were situations of risk, with the aim of finding out what was happening and weighing up possible interventions. It was in fact these 'reports' that often contained precious information on popular habits and orientations that today we would undoubtedly include in the field of research in medical anthropology, such as for example the widespread mistrust among the peasantry of the cures prescribed by official medicine.

1.2. *Between the eighteenth century and the very beginning of the nineteenth: the 'reports' to the 'enlightened' Courts and, above all, the 'Napoleonic statistics'*

The advent of the Enlightenment and the coming to the fore of the idea that, in order to govern a State, it was necessary to have a great amount of information available on the territory and its populations (indeed it is from the research and publication of this information that the original name 'statistic' comes), led to the various States in Italy formulating meaningful 'reports', in general drawn up at the request of the more 'enlightened' monarchies. Later, in the period of the so-called 'Napoleonic governments' in Italy, this meant extremely detailed and imposing surveys founded on very extended 'questionnaires', broken down into hundreds of items, all aimed at procuring information in the municipalities of each province, then reaching a synthesis for the entire State set-up. In this way an immeasurable quantity of socio-anthropological information was documented, much of it dealing with the situations, representations and practices of what today would be considered as being the field of medical anthropology. The following example may be taken as emblematic of the situation: the 1811 'statistical inquiry' carried out in the municipalities of the Kingdom of Naples, whose explicit aim was to calibrate the multiple initiatives then being undertaken (which we would now classify as 'health education'), brought to light the 'superstitions' that induced a wide strata of the population to refuse to submit to the great campaign then fully underway to vaccinate against smallpox.

1.3. *The paroemiological collections of the mid-nineteenth century*

The social and cultural changes and the political Restoration after the fall of the Napoleonic state set-ups and the revolutionary ideas that they spread throughout the whole of Europe, brought to a halt most of the preceding great surveys. Conversely, in the new romantic climate, in Italy as elsewhere, enquiries into language and 'popular literature' sprang up. It is in this climate that many surveys were carried out in various areas on popular proverbs, usually then classified according to the type of event of human existence to which they referred. There emerged an extremely rich folklore production regarding health and disease, medics and medicine, how to prevent infirmity, together with accounts on healthy nutrition in the various situations and different stages of life, and on madness, magical-religious protection, and death.

1.4. *The great (and ambivalent) era of positivism (1870–1920)*

When, in philosophy, as in all branches of science, positivism began to come to the fore between the last three decades of the nineteenth century and the First World War, in Italy as elsewhere in Europe, research topics that clearly relate to medical anthropology underwent a headlong development in all directions. This is, moreover, the context that gave birth to the disciplines of anthropology, to sociology and to psychology, and that was seeing great developments in biology and medicine.

The general framework is well-known, namely the demand for the eventual scientific exploration of all aspects of reality, and therefore *also* of man, of his 'natural history' and of his psychic features, all within the framework of a secular vision of the world and of life. The most striking discoveries, opening up the way to cognitive and operational implications, were at that time made in the field of biology, from the Darwinian conclusions regarding the evolution of species to the discovery of the micro-organisms responsible for a number of serious diseases. From this it appeared that it was precisely biology that constituted the solid and 'material' ground for research that was really scientific and free of any religious chains, research that was *also* inherent to man, to his behaviour and to human society. This was so much the case that, in this first phase of its development as a discipline, European anthropology – and the majority of Italian anthropologists too – were led to found their research methods and theoretical models on a substantially biological outlook. It was therefore in a biological key that they interpreted even the entire phenomenology of socio-historical processes and the dynamic itself of individual and collective subjectivity. This had heavy implications within 'social Darwinism', which ended in justifying an alleged superiority of the 'white race' over all others, and in justifying colonialism, the subordinate status of women, the repression of any form of deviance, and opposition to any form of public health finance for the needy strata of the population.

Despite these limits, yet while still being strongly influenced by them, research activities connected in various ways with what we now call the field of medical anthropology. They were quite substantial in this period and took three different directions:

(a) research into madness, its nature, its determinants and its possible somatic correlates, within an interpretative scheme in which strong analogies were outlined between mental disturbance, criminality, radical political dissent and in general all forms of deviance, all brought back to an overall common matrix of 'biological atavism';

(b) research into 'other' states of consciousness, in particular the hypnotic state, trances produced in the course of spiritistic séances, some forms of religious mysticism and even some manifestations of so-called 'collective psychology';

(c) research into representations and popular practices concerning 'negative influences' such as 'bad luck' or 'evil spells' and the defensive magico-religious responses bound up with them (magic formulae, amulets, and so on) and, with obvious relevance, research into folk medicines and into their operatives. All this research was conceived, on the one hand, as aiming at a precise documentation of the persistence of 'primitive' cultural forms in the most 'backward' strata. Those cultural forms were of great interest because they were held to indicate previous phases of human evolution that had now disappeared. On the other hand, this research was conceived as a necessary cognitive procedure aimed at highlighting ineffective and dangerous 'super-stitions' that had to be combatted, especially in the countryside. All this was done in the name of civil progress and with a secular vision of the world, in order to assert the value of academic medicine and its sole legitimacy, even in juridical terms.

In general these were projects carried out by single researchers, sometimes employed in universities or hospitals, but supported by a dense network of relations made possible by the recently born scientific societies and specialist periodicals. Here we may think of the great 'Enquiry into superstitions in Italy' (*Inchiesta sulle superstizioni in Italia*) launched in 1887 by the 'Archive for Anthropology and Ethnology', with the collaboration of many anthropologists through important field surveys in different parts of the country. But it was exactly in this period, which coincided with the processes of unification of the new Italian State, that the national Parliament undertook a number of imposing 'enquiries' into different aspects of society. And it is from these enquiries that important information emerges on the conditions of health of the popular strata, on the health resources available in the various territorial zones and also, for example, on the nutritional practices of the peasantry, or on the still active presence in each municipality of popular healers, i.e. men or women who 'exercise the art of health without being able to demonstrate they have a diploma from a university school of the Kingdom'.

1.5. *The two decades of fascism and the decline of the social sciences*

The subsequent period, in other words the two decades of the fascist era (1922–1945), is best characterized in the human sciences by the intersection of a number of processes:

(a) the general decay of the positivistic theoretical-methodological framework and its extreme sclerotization in physical anthropology with the emergence of the new idealist-historicist and spiritualist currents;

(b) the ensuing near-total abandonment of the methods and practice of empirical social scientific research into 'the present' and therefore the substantial disappearance of socio-anthropological surveys of the actual situation of the country; and instead

(c) a certain recovery of some lines of folklore studies, regarding especially the so-called 'popular arts' and ethnologic data collections in the colonies conquered in Africa, within a horizon stimulated by the new 'myths' proposed by fascism (the ancient values, intact within the peasantry, of the 'frugality' and 'harmony' of the 'Italian stock', together with the 'imperial civilizing vocation of the new Italy', heir to the power of ancient Rome).

This new ideological political context was characterized by the new emphasis placed on the primacy of biomedicine, one aspect of which was the great campaigns undertaken by the government to combat malaria and tuberculosis. In this climate, the previous widespread interest in folk medicine, which in any case had been bridled, now fell away and, to an even greater extent, so too did the interest in so-called superstitions. A few residual contributions to beliefs and practices of folk medicines remain, however, of note. Attention should also be drawn to the collections of 'ex voto' ('for grace received') offerings, namely artefacts offered on the return to health after accidents, illnesses etc. In addition, mention may be made of the forms of magic-religious protection adopted by Italian soldiers during the First World War and of a number of investigations into the medical beliefs and practices of the populations conquered in Africa during the war in Ethiopia (1935–1936).

2. The take-off and directions of development of Italian medical anthropology in the second half of the twentieth century

The attempt to outline the events and processes that marked the birth and developments of medical anthropology in Italy as an autonomous discipline, confronts the researcher with a fairly complex path, formed by the convergence of different lines and grounds of research and by non-homogeneous perspectives of work. Some of these are rooted in the old eighteenth-century studies of popular medicine, others represent strongly innovative research, stimulated by the general development of the psycho-social sciences and by the concrete problems posed by a rapidly evolving society and its health service. This is a path which, moreover, is partly connected to the reflections and empirical contributions produced in other countries, and partly specific to Italy. One may, for example, think of a number of conceptual and methodological models deriving from the writings of Antonio Gramsci and the founding value of the researches of Ernesto de Martino.

2.1. *Ernesto de Martino: the world of magic and the 'crisis of presence'*

In 1948 Ernesto de Martino published *Il mondo magico* (*The world of magic*), a fundamental text for the rebirth of both Italian anthropology and medical anthropology. The work's ethnographic background is provided by a thorough re-examination of the vast international documentation on magical rituality and on the states of consciousness found intertwined there, centred significantly on shamanism. His reading shows a substantial abandonment of Crocean idealist historicism, and a new interpretative opening towards cultural diversity and the critical acquisition of suggestions ranging from Pierre Janet through Gustav Jung to German phenomenological philosophy. It brings him to the following key questions. First there is the profound influence of psychic features, culturally constructed, on an individual's state of health. Second there is the historicity of forms of consciousness. Third even when 'negative' states of anxiety arise, the historical-cultural conditions nevertheless allow the constitution and maintenance of a *unitary* and substantially *autonomous* consciousness (the *Dasein* of German phenomenology, the 'being there' or 'presence'). Finally, there is magic rituality as a cultural device – even if archaic and linked to precarious conditions of existence – that is able in some way to 'guarantee' the 'presence'; that is to say it ensures the unity and autonomy of individual and collective consciousness through a subjective 'de-historification' of the negative situation. It reconfigures them in a meta-historical context in which it had already been 'resolved', namely in 'mythical time'. This is a subject on which de Martino was to continue to work for the whole of his life. It was central in the 1958 volume *Morte e pianto rituale nel mondo antico* (*Death and ritual lamentation in the Ancient World*) and in *Sud e magia* (*The south and magic*). This subject was to return in his *Introduction* to the collection of classic texts in ethnology and historical-religious studies, *Magia e civiltà* (*Magic and civilization*) that he edited (de Martino 1962), right up to *La fine del mondo* (*The end of the world*), a posthumous publication edited by Clara Gallini (de Martino 1977). The subject of this latter volume is however to be seen in the 1964 essay 'Apocalissi culturali e apocalissi psicopatologiche' (*Cultural Apocalypses and Psychopathological Apocalypses*) published in *Nuovi Argomenti*, the journal then edited by Alberto Moravia and Alberto Carocci (de Martino 1964).

2.2. *Medical anthropology's first contributions to health strategies*

It is really to the second half of the 1950s that, for Italy too, we can date the birth of modern medical anthropology.

1956 saw the publication of the author's critical essay on the use of ethnological research in the WHO's country-specific strategies in the so-called Third World (Seppilli 1956). This was the first time the author had written on medical anthropology. In fact, it appears to have been the first Italian contribution entirely and specifically carried out on the very terrain of modern medical anthropology.

Shortly thereafter, the author's scientific, teaching and operational collaboration as an anthropologist began with the Experimental Centre for Health Education of the University of Perugia. This still ongoing collaboration, which began in 1958 with a lecture published shortly afterwards (Seppilli 1959), was to be enlarged to numerous projects of both research and intervention, and to holding courses of cultural anthropology and sociology in the School of Specialization in Hygiene and Preventive Medicine. Finally, it resulted in the master's and doctoral programme in Health Education, again at the University of Perugia.

2.3. *Research into popular medicine*

1957 marks Ernesto de Martino's last, and most detailed and important, 'expedition' into Lucania, where he carried out his exemplary research on healers and their patients. In the course of this field research, he explored the question of the structure of magico-therapeutic formulas (the so-called *historiolae*) and the mythico-ritual reference scheme of exorcisms in general, as well as the exorcisms' cultural function. The research group also included a psychoanalyst (Emilio Servadio, of the Parapsychology Foundation which supported the initiative) and a medic (Mario Pitzurra), who was already working with the author's group in Perugia, as also was the head of the photographic documentation team, Ando Gilardi. The photographic documentation for the research constituted the object of the First Italian Exhibition of Ethnographic and Sociological Photography held in 1958, first in Perugia, then in Rome.

Again in 1957, with the Institute of Anthropology that the author had recently established in the University of Perugia, the author began a series of enquiries into popular medicine and peasant healers. His focus was on their training, their ideological-cultural horizons, their activity and their clients, in Umbria and in other regions of central Italy (Seppilli 1983, 1989). In the course of these enquiries, which went on until the 1990s, there emerged a strong psychic component, right from the start, in the healers' therapeutic processes, which was particularly effective when dealing with a traditional syndrome emically known as the *'fattura'*. Here one must underline a specific quality on the relation between healer and patient, that is to say the sharing of cultural and linguistic references relating to 'ills' and the request that the patient participate in the diagnostic and therapeutic processes. In the context of this line of research into popular medicine, attention should be drawn to: (a) the study, restoration and cataloguing of one of the most important European collections of ancient and contemporary amulets, gathered by Giuseppe Bellucci between 1870 and 1920; (b) the research into votive cults and the cataloguing of ex-voto offerings located in different sanctuaries in central Italy; (c) research focused on specific 'ills' as classified by popular medicine; (d) the historical reconstruction of the relations between popular medicine and official medicine; in the Italy of the nineteenth and twentieth centuries; (e) the theoretical contributions to an anthropological definition of 'popular medicine' within the context of the wider debate, of Gramscian origin, on the significance of 'popular culture', on its nature, on its social bases and on its relations with the systems of hegemony and power.

In particular from the 1970s onwards, the research on healers and popular medicine, and also on pilgrimages and votive deposits, became wide-ranging, involving numerous anthropologists and other scholars in many regions in Italy, who gave it notable weight and significance.

2.4. *The question of tarantism and the birth of Italian ethnopsychiatry*

In 1959 Ernesto de Martino completed his by now classic research on *tarantism* in the Salento district of Puglia (Apulia). The research was carried out as ethnographic 'field work' but at the same time it was a careful historical reconstruction of a great quantity of news items and information that in the course of previous centuries had been produced around this veritable 'culture-bound syndrome'. The research confronted once more the question of the relation between disease 'in the medical sense' and how the disease is experienced, shown and ritually faced. The team, which also produced an ample visual and audio documentation, included an ethnomusicologist (Diego Carpitella), a psychologist (Letizia Jervis-Comba) and a psychiatrist (Giovanni Jervis). The outcome was perhaps the best-known of de Martino's works, *La terra del rimorso* (*The land of remorse*) (de Martino 1961), a volume that marks the constitution in Italy of ethnopsychiatry, which was subsequently to develop in different directions. A short time afterwards, a collaborator of de Martino's, Clara Gallini, undertook research in Sardinia on a phenomenon, *argia*, that in many ways is similar to *tarantism*. In this case too, the same ethnomusicologist (Diego Carpitella) and two psychiatrists (Giovanni Jervis, again, and Michele Risso) took part in the research.

From this line of research and the successive convergence, on different occasions, of the expertise provided by anthropology and by psychiatry, Italian ethnopsychiatry got underway. It was initially oriented towards: (a) 'internal' processes that were connected predominantly with migratory phenomena from the country's South to the North and thence as far as Switzerland and then, more recently, migratory flows into Italy from a great number of other countries in Eastern Europe, Asia, South America and Africa; and (b) mental disturbance and the so-called traditional responses that come to the fore in other continents, in particular in Africa, but also in Asia and South America.

2.5. *Medical anthropology and the fight against the asylums*

Starting from the mid-1960s, the relationship between anthropology and psychiatry began to develop, especially in Umbria, on another terrain too. This concerned the anthropological participation in the research and the combat strategies of the great anti-institutional movement that led in 1978 to the abolition of the asylums and the construction of territorial psychiatric networks. To the present-day this debate has continued, on questions regarding both the experiences and perspectives of these services and also the definition of a more general policy for mental health.

2.6. *The great development*

Medical anthropology in Italy is by now in full development. From the beginning of the 1980s, there has been a convergence of the various lines of research in an overall disciplinary corpus under the common denominator of 'medical anthropology'. In 1983 a first, substantially unitary conference took place in Pesaro, in 1988 the Italian Society of Medical Anthropology (SIAM) was formed in Perugia (Seppilli 1996) and in 1996 publication began of its review periodical '*AM. Rivista della Società italiana di antropologia medica*'. It was in this period that many regions developed both teaching and an operational link with the National Health Service (SSN) and that medical anthropology began to be offered first in the humanistic faculties and then in the medical ones.

3. A first synthesis

It seems appropriate here, in conclusion, to attempt to synthesize what seem to be the specificities, or at least the main defining characteristics of Italian medical anthropology.

- Permanent attention is paid to the general set-up of the human sciences and to the relationship of both reciprocal autonomy and overall integration between the approaches of a biological type and the socio-historical ones.
- Within this systemic horizon, a 'particular' relationship with biomedicine is entertained: posed in certain ways as a simple emic object of research – one of the many medical systems in human history – and in other ways as an interlocutor with whom in some way the scientific conception of the world and, in particular, the scientific approach to the phenomenology of health/illness are shared epistemologically. It is within this perspective that the anthropological critique is situated of biomedicine being insufficiently scientific since, to a large extent, it excludes subjectivity and the ensemble of social determinations from its cognitive and operative set-up.
- Again within the systemic horizon – and also within the perspective of a possible interpretative path of the 'effectiveness' of the various medicines – great attention is given to the importance of psychic features, as a culturally conditioned lived experience, over the entire arc of the processes of health/illness, of the practices of healing and in general of 'taking care' of oneself: from this stems the privileged relationship that medical anthropology has with psychoneuroendocrinoimmunology.
- Equally, great attention is given to the social determinants of the processes of health/illness, to social inequalities in risk factors and in access to the services, to the functioning of the structures of the health services, with a consequent strong emphasis placed on the need for awareness in taking part in political choices regarding health (health as a 'common good').
- On a great part of the subjects dealt with, there is a detailed involvement as regards teaching in the training and updating and refresher courses of medics and other social health workers and in the formulation of the strategy of the health services. It is exactly in this perspective that there is currently a converging pressure – for a greater weight of the 'social factor' in the training curricula of the faculties of medicine – coming from the five

Italian scientific societies that have the specific task of bringing together researchers working in the various disciplinary fields of the social sciences whose scope is health (medical anthropology, psychology of health, sociology of health, health economics and history of medicine).

- Italian medical anthropology is broadly concentrated on 'domestic' processes; research 'abroad' is almost all fairly recent and in part conditioned by the framework of so-called 'international cooperation'; despite a recent development in this direction of ethnopsychiatric research, as we have seen this field too had its beginning 'at home' in Italy in 1959 and today has developments that are linked in a particular fashion with recent immigration.

- There is a theoretical-methodological approach on which there has been a strong influence of the interpretative framework developed by Antonio Gramsci in regard to structures of power and mediation and to the processes of hegemony and the circulation of culture. In essence, we may indicate general perspectives, thematic options and institutional reference points oriented towards a marked socio-political engagement.

Acknowledgements

Translation from the Italian by Derek Boothman. The paper was presented at the conference 'Medical Anthropology in Europe' funded by the Wellcome Trust and Royal Anthropological Institute.

Conflict of interest: none.

References

de Martino, E. 1948. *Il mondo magico. Prolegomeni a una storia del magismo.* Torino: Einaudi.
de Martino, E. 1958. *Morte e pianto rituale nel mondo antico. Dal lamento pagano al pianto di Maria.* Torino: Einaudi Boringhieri.
de Martino, E. 1959. *Sud e magia.* Milano: Feltrinelli.
de Martino, E. 1961. *La terra del rimorso. Contributi a una storia religiosa del Sud.* Milano: Il Saggiatore.
de Martino, E., ed. 1962. *Magia e civiltà.* Milano: Garzanti.
de Martino, E. 1964. Apocalissi culturali e apocalissi psicopatologiche. *Nuovi Argomenti* 69–71: 105–41.
de Martino, E. 1977. La fine del mondo. In *Contributo all'analisi delle apocalissi culturali,* ed. C. Gallini, Torino: Einaudi.
Seppilli, T. 1956. Contributo alla formulazione dei rapporti tra prassi igienico-sanitaria ed etnologia. In *Atti della XLV Riunione della Società Italiana per il Progresso delle Scienze (Napoli, 16–20 ottobre 1954),* Vol. II, 295–312. Roma: SIPS.
Seppilli, T. 1959. Il contributo della antropologia culturale alla educazione sanitaria. *L'Educazione Sanitaria* IV, no. 3–4: 325–40.
Seppilli, T., ed. 1983. La medicina popolare in Italia. *La Ricerca Folklorica,* 8: 3-143.
Seppilli, T., ed. 1989. *Le tradizioni popolari in Italia. Medicine e magie.* Milano: Electa.
Seppilli, T. 1996. La 'Società italiana di antropologia medica' (SIAM). *AM. Rivista della Società italiana di antropologia medica* 1–2: 361–66.

Before and after fieldwork: ingredients for an ethnography of illness

Gilbert Lewis

University of Cambridge, Cambridge, UK

W.H.R. Rivers asked what light anthropology could throw on the emergence of medicine. But this early lead was not soon followed up. Social anthropology had first to establish itself. The new social anthropology championed holistic fieldwork in small-scale societies. Some did choose to study illness or misfortune ethnographically. Evans-Pritchard, Victor Turner and Max Marwick provided outstanding models in this field. Political change and decolonisation made some of the older assumptions about the place and aims of anthropological research less easy to sustain. Growth in the subject encouraged specialisation. Medical anthropology was one among many possible developments. But it also had to identify a distinctive focus and place in relation to medicine and other health-related social studies.

Rivers

It might seem odd that medical anthropology did not emerge sooner – medical men (e.g. Hunt, Rivers, Broca, Seligman) had been prominent in the first anthropology societies. In Britain, W.H.R. Rivers was the exception and a shaping influence. His FitzPatrick lectures in 1916 were on *Medicine, magic and religion* (Rivers 1924); his question was: What light does anthropology throw on the emergence of medicine as a social institution? He examined Melanesian concepts of disease and treatment practices in relation to ideas of cause, magic and religion. His approach was sociological and comparative. Later commentators have applauded much in his discussion (for example, about causal reasoning, the coherence in Melanesian ideas and practice, the part played by suggestion in illness). But some critics have thought diffusionism tainted his approach in the second half[1] – his stress on 'transmission as an important factor in interactions between peoples'. There is some irony in this given that so much work in recent medical anthropology has been on borrowings, the adoption of exotic or alternative medicine, pluralism, responses to introduced western medicine, the effects of introduced diseases.

There was no quick uptake of his ideas. His friend and literary executor G. Elliot Smith did imagine creating at University College London (UCL) in the Anatomy

Department (where Elliot Smith was professor) a centre for anthropology, psychiatry and comparative neurology: he wrote about it in a letter dated January 1926 (Dawson 1938, 89–95). It was an early but unrealised vision, a vision, almost, of things to come.

Social anthropology – its emergent characteristics

Instead, a shift from Ethnology to Social Anthropology marked the period following Rivers. Ethnography and intensive fieldwork came to characterise what was special about British social anthropology – with the brilliant example of Malinowski's fieldwork. It, like Rivers's lectures, had taken place during the First World War. The emphasis was on fieldwork in an 'other' culture. The method was firsthand observation in a living society, using the local language, learning the insiders' views. The 'totalising' approach aimed at study of the whole so as to see connections (between persons, events, institutions, rules, ideas, rites and practices). This fieldwork method lay at the heart of the (structural-) functionalism which became its theory.

And it was fieldwork in a certain type of society – small-scale, preliterate, relatively undifferentiated, with simple technology. In theory, anthropology was about all humankind; in practice about the small scale and the exotic. There was a kind of justification for this; in the small-scale society it was more practicable, more possible to see and study the whole. Anthropology was different – not sociology: it was not done at home. The 'exotic' had the attraction of discovery, the challenge of finding something new and strange – something different from what might be held to be normal, or natural, or necessary at home. Social anthropology was comparative and relativistic. It could call into question what was taken for granted.

Productive tensions developed – conflicting pulls between opposing poles of attraction (Freedman 1978, 8–32): the outsider's view or the insider's view; 'objective' or 'subjective'; a focus on society or on culture; on description or on theory; comparison or particularism; was the aim to be science (a social science) or to be one of the humanities?; to study the 'traditional' or the present-day. Maurice Freedman (1978, 134) noted in a searching review for Unesco of the state of anthropology: 'Yet behind the heterogeneity there is a basic programme: the pursuit of "totality" – not the study of everything pertaining to man but the "total" study of whatever it is that is chosen for investigation. If anthropology had confined itself to the study of the small-scale and the non-literate, it would have had a clearer profile, but at the cost of intellectual adventure'.

As to the 'totalising' field method carried out by a single investigator, it was, of course, self-limiting. Social anthropology was by nature interdisciplinary. But no one can really master everything, be truly jack-of-all trades. Inevitably people began to specialise on an aspect of the whole – kinship or law, for example. But if someone specialised, how could they take account of the established discipline in the chosen field? Economics, say? Or medicine? Was it accessible? There were rare examples such as that of Margaret Field, government anthropologist in the Gold Coast in the 1930s, whose observations of social change, anxiety and illness led her to return to Britain, train in medicine then in psychiatry, and go back in 1955 to study mental illness among people attending Akan shrines (Field 1960). The Medical Research Council (MRC) funded her for that research.

To specialise or collaborate in research?

Anthropological findings did attract medical interest. For example doctors with experience of working abroad saw its relevance for practice and for teaching. The MRC wanted to explore possible areas of research collaboration – in public health and tropical medicine, for example. But at that time, the 1950s, the main anthropologists approached were reserved in response (Firth 1978)[2] – there were so few professional anthropologists, and they were preoccupied with establishing their own subject first. Was medicine to be master and anthropology servant?

Psychiatrists also looked at anthropology for its possible bearing on questions of diagnosis, social change and mental illness. Kiev, then at the Institute of Psychiatry in London, edited a volume (Kiev 1964) on magic, faith, and healing (with articles by V. Turner, R. Prince, J. Dawson and others). The Ciba Foundation organised a symposium in London on Transcultural Psychiatry (de Rueck and Porter 1965). A galaxy of authorities from both fields attended it: Firth, Fortes, Margaret Mead, Hallowell, Leighton, Lambo, Yap, Murphy, De Vos, Carstairs. Such meetings brought the subjects together and helped to identify interests in common. Meyer Fortes and Joe Loudon, who were both at the Ciba meeting, were the prime movers in organising the ASA 1972 Meeting on Social Anthropology and Medicine.

Psychiatry and anthropology have certain overlapping interests. Both tackle problems of cultural interpretation and understanding others. Apart from Rivers and Field, others had contributed to transcultural psychiatry before it became a well-recognised subject: they included anthropologists (G. Bateson, A.P. Elkin, G. Roheim, E. Goffman, F.E. Williams) and psychiatrists (J.C. Carothers, M. Carstairs, T.A. Lambo, J.B. Loudon, H.B.M. Murphy, J. Orley, R. Prince, C.G. Seligman, S.M. Shirokogoroff, G. Tooth, P.M. Yap). The work of Durkheim on suicide, Kraepelin's concern with diagnosis and nosology, Freud and Janet on hysteria, automatisms and dissociation, had also helped earlier to establish intellectual foundations for a sociology of mental disorder and the cross-cultural study of spirit possession and shamanism. Certain French initiatives were notable, especially those of R. Bastide, and of H. Collomb and the Dakar group of doctors and anthropologists associated with Collomb's journal *Psychopathologie africaine* (among them A. Zempléni). Influences from Foucault's writings and the anti-psychiatry movement came later. Ellenberger's *History of the unconscious* appeared in 1970. It was on the evolution of dynamic psychiatry and its significant originators. Possession, suggestion and faith healing were brilliantly explored in it as well as relationships between healer and patient. It began with 'The study of primitive healing ... of interest not only to anthropologists and historians (as being the root from which, after a long evolution, psychotherapy developed) but ... also of great theoretical importance to the study of psychiatry as the basis of a new science of comparative psychotherapy' (Ellenberger 1970: 3).

In the 1960s in Britain, there was no textbook on medical anthropology. What came nearest was Susser and Watson's (1962) *Sociology in medicine*. It had a decidedly anthropological approach (with examples about tuberculosis in South African miners and illness in the context of the developmental cycle of domestic groups). Both authors taught at the University of Manchester, and both had previously worked in Southern Africa, one as a doctor, the other as an anthropologist.[3] Subsequently, the Royal Commission on Medical Education 1968 (The Todd Report) included sociology *and* social anthropology as disciplines useful for training medical scientists. But as Loudon (1976, 19–23) noted, the real problem was identifying when and how their introduction might be effective in that training.

Choosing a subject for research

Real engagement in anthropology required fieldwork, the qualifying experience. In the 1960s that meant fieldwork abroad. The question was what topic? There were already examples in anthropology of what might be done on illness and misfortune. Three of the outstanding ones were: Evans-Pritchard's (1937) *Witchcraft, oracles and magic among the Azande* – on reasoning about cause and responses to misfortune within a different framework of ideas; Turner's (1957, 1963, 1964) Ndembu studies, with his deployment of the extended case method revealing people's commitments to values and analysing their use of symbols and ideas in ritual and healing; and Marwick's (1961) *Sorcery in its social setting* with his systematic collection of data, the pool or universe of cases, the functionalist analysis of sorcery accusation as 'social strain gauge'.

In fact a number of people, roughly about this time (1960s), had chosen independently to focus on illness and treatment: Jean Buxton (1973) wrote on Mandari shrines and spiritual healing, their understanding of disease and the body and colour symbolism, from fieldwork begun in the 1950s although not published until 1973; Murray Last researched illness among Hausa villagers, the household, age and gender differences, research intended also to provide information useful for teaching medical students at Ahmadu Bello University, Zaria. Harriet Ngubane, herself Zulu, did research on the body and mind in Zulu medicine. Ronald Frankenberg and Joyce Leeson studied healers and theories of disease and misfortune in Lusaka, Zambia; Una Maclean, the complexity of Yoruba sickness behaviours in Ibadan, a huge city; Vieda Skultans, spiritualist healing in South Wales; Gilbert Lewis, Sepik New Guinea village concepts of illness; Eva Gillies, levels of causal explanation and discriminations of different kinds of illness made by Nigerian Ogori. Results from some of these researches were presented at the ASA conference held in 1972 (Loudon 1976).

A second theme, prominent at this time, already had a long history – healing in relation to religion and magic. Impressive new work included I.M. Lewis's (1971) provocative analysis of spirit possession and shamanism, Bryan Wilson's (1973) sociological study of new religious movements among colonised peoples (many of these focussed on faith and healing), and J.D.Y. Peel's (1968) field-based and historical study of prayer and healing in Aladura and Apostolic churches in Nigeria.

Some anthropologists signalled interest in the body. Mary Douglas (1966) had just published *Purity and danger*, her study of the body, pollution and taboo. Although not a central concern of anthropologists then, it would later stimulate a great growth of interest and subsequent work on the body and symbolism, on notions of normality and nature.

It might be asked: why take 'illness', then, rather than 'health', or 'the body' or 'medicine' to identify 'medical anthropology'? One could argue that 'health' or the 'body' are just as much of universal interest as 'illness' but even more neglected. Perhaps it is because medicine in origin depends on recognition of a change in self or someone. People may take health largely for granted until something hurts or fails. It is the change – the 'illness' – that must be explained, that calls for a response.[4]

The question of justification for the research

Research is not just a matter of what seems interesting or worthwhile to the investigator (the 'scientist') choosing what to investigate and where (Barnes 1979). There are also questions of acceptability, competence and funding. Social research must be justified to others: to the subjects of the research (the 'citizens' who may or may not agree to take part), to the 'sponsors' (who give grants, fund the research), to the 'gatekeepers' (who grant permission – government authorities, an ethics committee, local councils). Each will have particular expectations. The investigator will incur obligations.

In the author's own case, the MRC initially had funded him as a medic to get some grounding in social anthropology (perhaps for future medical research). But seeing his research proposal for an ethnography of illness in a New Guinea village, the MRC suggested he apply to the Social Science Research Council, who funded the research for a PhD in social anthropology. The local government authorities in New Guinea gave him permission, plus generous medical supplies, seeing the research perhaps as something that might be relevant to public health care. The villagers accepted him, despite inept attempts at explaining his plans, without declaring their own rather different hopes (in which he was bound to disappoint them).

For funding agencies, as also for those seeking research support, there were questions of research focus, potential career direction, what the research might contribute, and whether to theory or to practice? Social anthropology was a small academic subject, but medicine a long-established profession, large and highly diverse. The opportunities were unclear.

In the field

The research question was: What was it like to be ill in that sort of society? The aim was to include all illness whether explained or not, treated or not (Lewis 1975). To see what illness they faced both in their terms and in external medical terms. And to investigate what their explanations depended on. How did the outcome of an illness reflect back on people's ideas about its cause, or about the efficacy of a treatment, or the validity of an explanation? Partly the models for this were epidemiological – to follow a small, defined population over time, to see the range of problems and what they did about them.

In the field there was the practical question – what does participant observation really mean? Should one intervene (or interfere) or not, especially in a study of illness? Intervening, sometimes, certainly did bring closer access to people and reveal things about their sickness behaviour and their responses to it (and also those of the inquirer). Clinical observation may be concentrated too narrowly on the sick person. People are affected by expectations, by who is present, whether to hide or reveal illness. Mechanic (1962, 1968) showed that it was plainly a mistake to consider only the patient in isolation and not also the situation, and others around the patient.

Then there was the importance of language in illness: with a tape recorder one could go back over talk and events in detail, even if one could not control their timing and was only a beginner in their language. Frake (1961) was the inspiration here – the diagnostic criteria of local people revealed by listening to their questions and talk, not by asking them leading questions.

Writing up

Anthropological fieldwork rarely has a clear endpoint other than the practical one of a time-limit. And writing up must follow: 'Ethnography and theory are sometimes combined (as Clifford Geertz once remarked) like the ingredients of elephant and rabbit stew; one elephant of ethnography to one rabbit of theory. It is a good recipe for a cook who can bring out the flavour of the rabbit' (Lienhardt 1985, 647).

At the time, there seemed to be a number of issues to address. There was the question of selective attention. Evans-Pritchard had combined witchcraft, oracles and magic because they formed a system of reciprocally dependent ideas and practices. But that decision left in shadow certain other Zande responses to illness (Gillies 1976). There is a temptation to focus chiefly on exceptional cases or on the dramatic. To counter the problem of selective attention, one could take the 'pool' or 'universe' of illness and consider all the responses – not just the interesting ones or the witchcraft or sorcery cases.

Then there were issues about illness explanations: what role might illness or treatment have in changing prevailing ideas? Nadel (1953, 175) had raised this with a question about a moral system threatened by the fact that, with new drugs, a supernatural punishment can be made to disappear over night. How did people cope when illness or its outcome challenged what was promised or expected? E.H. Carr's analysis (Carr 1961: chapter 4) of thinking about the causes of an accident was helpful by analogy for understanding Sepik approaches to illness causation, their efforts to account for the particular, for why someone was singled out for illness. Illness prompted them to review recent events. It revealed (more broadly) some of their views on their world, the forces present in it, and human nature.

Developments

Changes were of course affecting social anthropology: political independence and social changes altered people, places and the possibilities for fieldwork. They were forcing social anthropology to redefine its identity and aims. Academic departments and students of anthropology had grown in numbers and were branching out. Specialisms began to emerge.

Change hit social anthropology, but there were changes affecting medicine too. McKeown (1976) sought to dethrone the view that progress in medicine was the reason for the modern rise in population. But the impact of antibiotics on infectious diseases did alter perceptions of the patterns and problems of disease as well as frameworks of explanation in medicine (Kunitz 1983, 179–87; 1987). The spotlight shifted more onto chronic and degenerative disease, where specific causes and specific treatments are not so clear. Chronic diseases came to dominate the illness profile that doctors see, and they complained about problems they considered really social problems – but that is the complex reality of their work.

Fabrega (1972) surveyed the literature on medical anthropology, producing an impressive guide to the extremely scattered field. He proposed a framework for its study. His focus was disease in social life and the comparative study of illness behaviour – rather than medical systems as such. But a focus on medical systems was to become increasingly prominent, especially via critiques of the dominant medicine – western biomedicine – in its US and European mainstream versions as well as in other settings, and in relation to alternative kinds of medical practice. In part, this derived from the sociology of medicine initiated by Talcott Parsons in *The social system* (Parsons 1951, chapter 10). Eliot Freidson's *Profession of medicine* (1970) strengthened it. The symposium on *Asian medical systems* edited by Charles Leslie (1976) heralded social anthropological research on other medical systems and on medical pluralism. There was, in effect, a marked turn towards the sociology/social anthropology *of* medicine rather than sociology/social anthropology *in* medicine.

In 1976, the Medical Sociology Group of the British Sociological Association organised a meeting 'to stimulate a real dialogue between medical sociology and social anthropology' (Firth 1978). In fact, 20 years before, Margot Jeffreys, then sociologist at the London School of Hygiene and Tropical Medicine, had opened Raymond Firth's 1956 LSE seminar on medical anthropology with a paper identifying the potential areas for research collaboration. Many later examples could show overlaps of theory and interests in medical sociology and anthropology: for example, in concepts of health, gender and caring roles (Currer and Stacey 1986; Stacey 1988), blood donation and the notion of the gift (Titmuss 1970), the social origins of depression (Brown and Harris 1978; Littlewood and Lipsedge 1982), generational differences in experience of illness and health services (Blaxter and Patterson 1982), medical systems viewed as socio-cultural systems (Comaroff 1978; Worsley 1982).

Recognition as a subject

In 1976, BMAS (British Medical Anthropology Society) was founded by Cecil Helman and Joseph Kaufert as a forum for medical anthropology. It was unstructured, it had no officers, and no money. From the outset it was intended to provide a common ground for the exploration of medical anthropology research, ideas and applications. It took the form of meetings to which anyone interested could come. As proposed in its original statement: 'Medical anthropology is neither a closed shop nor a specialised offshoot of social anthropology – nor a collection of medical exotica gathered by doctors who happen to have worked abroad'. There was a loosely structured steering group. The meetings depended on people's interest and initiative, on someone undertaking to organise speakers and venue. Over the first ten years, almost 40 meetings were held at 23 different venues in England, Scotland and Wales. Participants included – as well as anthropologists – psychiatrists, social workers, public health physicians, nurses, GPs, sociologists and medical historians. BMAS (which still exists) produced a newsletter, which ran from 1981 until 1993 when this journal, *Anthropology and Medicine*, replaced the newsletter.

In the 1970s optional courses in medical anthropology began to be offered to undergraduates in the social anthropology departments at London, Cambridge, Sussex and Keele. UCL started its Centre for Medical Anthropology (under Murray Last and Roland Littlewood) and held a weekly research seminar attracting speakers and audience from a wide range of fields and cognate interests. In 1978, the RAI recognised a growing topic with the biennial award of a Wellcome medal[5] for 'research in anthropology as applied to medical problems'.

Teaching in medical anthropology was also established at Brunel, the London School of Hygiene and Tropical Medicine, the School of Oriental and African Studies, Goldsmiths and UCH medical school. In the 1980s this was increasingly developed at postgraduate course level both for those with health work experience and for those with social science qualifications whose interest was now particularly in the biomedical application of anthropology. Cecil Helman, who taught at Brunel and UCH Medical School, was especially concerned to introduce medical anthropology to health professionals.

Acknowledgements

The author thanks the editors, the anonymous referees, Stephen Kunitz, Murray Last and Jerome Lewis for helpful suggestions.

The paper was presented at the conference 'Medical Anthropology in Europe' funded by the Wellcome Trust and Royal Anthropological Institute.

Conflict of Interest: none.

Notes

1. Rivers latterly became influenced by Perry and Elliot Smith's views on the spread of cultural traits by diffusion from an original centre – their 'hyperdiffusionist' version proposed ancient Egypt as the original source of civilisation (Rivers 1924, vi–vii, 90,106; Firth 1978, 238).
2. For three years from 1956, Raymond Firth chaired a seminar at the London School of Economics on medical anthropology, seeking partly to make clearer what role there might be for anthropology in relation to public health.
3. The contribution to medical anthropology of people with South African experience is striking: J. Cassell, Jean Comaroff, Meyer Fortes, Cecil Helman, M. Gelfand, S.L. Kark, Joe Loudon, Harriet Ngubane, M.W. Susser.
4. 'Health' is the unmarked member of a linguistic pair. 'Health' may include 'illness' (e.g. 'health statistics', the question 'How's your health?') but not the reverse – 'illness' does not include 'health'. That markedness fits the way people take health for granted, but illness makes them stop and think.
5. In 1931, Sir Henry Wellcome instituted an annual medal for 'applied anthropology'. It continued until 1966 then lapsed. In 1977 the Wellcome Trust agreed with the Royal Anthropological Institute to revive it biennially, but with narrowed scope.

References

Barnes, J.A. 1979. *Who should know what?: Social science, privacy and ethics*. Harmondsworth: Penguin.

Blaxter, M., and E. Patterson. 1982. *Mothers and daughters: A three-generational study of health attitudes and behaviours*. London: Heinemann.

Brown, G.W., and T. Harris. 1978. *Social origins of depression*. London: Tavistock.

Buxton, Jean. 1973. *Religion and healing in Mandari*. Oxford: Clarendon Press.

Carr, E.H. 1961. *What is history?* Harmondsworth: Penguin.

Comaroff, Jean. 1978. Medicine and culture: Some anthropological perspectives. *Social Science and Medicine* 12B: 247–54.

Currer, C., and M. Stacey, eds. 1986. *Concepts of health, illness and disease*. Leamington Spa: Berg.

Dawson, Warren R., ed. 1938. *Sir Grafton Elliot Smith: A biographical record by his colleagues*. London: Jonathan Cape.

De Rueck, A.V.S., and R. Porter, eds. 1965. *Transcultural psychiatry*. London: Churchill.

Douglas, Mary. 1966. *Purity and danger*. London: Routledge and Kegan Paul.

Evans-Pritchard, E. 1937. *Witchcraft, oracles and magic among the Azande*. Oxford: Clarendon.

Ellenberger, Henri F. 1970. *The discovery of the unconscious: The history and evolution of dynamic psychiatry*. London: Allen Lane The Penguin Press.

Fabrega, H. 1972. Medical anthropology. In *Biennial review of anthropology 1971*, ed. B.J. Siegel, 167–229. Stanford: Stanford University Press.

Field, Margaret. 1960. *Search for security: An ethno-psychiatric study of rural Ghana*. London: Faber and Faber.

Firth, Rosemary. 1978. Social anthropology and medicine – a personal perspective. *Social Science and Medicine* 12B: 237–45.

Freidson, Eliot. 1970. *Profession of medicine*. New York: Dodd, Mead.

Frake, C.O. 1961. The diagnosis of disease among the Subanun of Mindanao. *American Anthropologist* 63: 113–32.

Freedman, Maurice. 1978. Social and cultural anthropology. In *Main trends of research in the social and human sciences*, ed. J. Havet, Vol. I. 3–177. The Hague/Paris: Mouton/Unesco.

Gillies, Eva. 1976. Causal criteria in African classifications of disease. In *Social anthropology and medicine*, ed. J.B. Loudon, 358–95. London: Academic Press.

Kiev, Ari. 1964. *Magic, faith and healing*. New York: Free Press.

Kunitz, S.J. 1983. *Disease change and the role of medicine: The Navajo experience*. Berkeley: University of California Press.

Kunitz, S.J. 1987. Explanations and ideologies of mortality patterns. *Population and Development Review* 13, no. 3: 379–408.

Leslie, Charles, ed. 1976. *Asian medical systems: A comparative study*. Berkeley: University of California Press.

Lewis, Gilbert. 1975. *Knowledge of illness in a Sepik society*. London: Athlone Press.

Lewis, I.M. 1971. *Ecstatic religion: An anthropological study of spirit possession and shamanism*. Harmondsworth: Penguin.

Lienhardt, Godfrey. 1985. Review of G. Stocking, ed., *History of anthropology*. *Times Literary Supplement*, June 7 1985, p. 647.

Littlewood, R., and M. Lipsedge. 1982. *Aliens and alienists*. Harmondsworth: Penguin.

Loudon, J.B., ed. 1976. *Social anthropology and medicine*. London: Academic Press.

Marwick, M.G. 1961. *Sorcery in its social setting*. Manchester: Manchester University Press.

McKeown, T. 1976. *The modern rise of population*. New York: Academic Press.

Mechanic, D. 1962. The concept of illness behavior. *Journal of Chronic Diseases* 15: 189–94.

Mechanic, D. 1968. *Medical Sociology*. New York: Free Press.

Nadel, S.F. 1953. Comment. In *An appraisal of anthropology today*, eds. S. Tax, L.C. Eiseley, I. Rouse, and C.F. Voegelin, 175. Chicago: University of Chicago Press.

Parsons, Talcott. 1951. *The social system*. New York: Free Press.

Peel, J.D.Y. 1968. *Aladura*. Oxford: Oxford University Press for the International African Institute.

Rivers, W.H.R. 1924. *Medicine, magic, and religion*. London: Kegan Paul, Trench, Trubner.

Stacey, Margaret. 1988. *The sociology of health and healing*. London: Unwin and Hyman.

Susser, M.W., and W. Watson. 1962. *Sociology in medicine*. Oxford: Oxford University Press.

Titmuss, R.M. 1970. *The gift relationship: From human blood to social policy*. London: Allen and Unwin.

Turner, V.W. 1957. *Schism and continuity in an African society*. Manchester: Manchester University Press.

Turner, V.W. 1963. *Lunda medicine and the treatment of disease*. Rhodes Livingstone Papers No. 15. Lusaka: Govt. Printer.

Turner, V.W. 1964. An Ndembu doctor in practice. In *Magic, faith and healing*, ed. A. Kiev, 230–63. New York: Free Press.

Wilson, Bryan. 1973. *Magic and the millennium*. London: Heinemann.

Worsley, Peter. 1982. Non-Western medical systems. *Annual Review of Anthropology* 11: 315–48.

Vignette

The beginnings of medical anthropology in France

Questions concerning the cultural and social dimensions of illness had been examined in several works well before a 'medical anthropology' came into existence in France. This happened in two currents of research: one of the physicians who had been confronted with society and culture in medical practice, the other of philosophers and anthropologists interested in the social and the cultural aspects of disease and its treatment. Among the first, two researchers should be remembered: Charles Nicolle, for insisting on the social and historical aspects of infectious disease, and especially René Leriche, who already in the 1950s proposed to draw a distinction between the 'doctor's disease', object of medical scientific research, and the 'patient's illness', experienced and interpreted in a social context. Among the works of the philosophers and anthropologists, texts by Claude Lévi-Strauss, published in 1949, played a pioneering role. Lévi-Strauss discussed questions of religion, but in a fashion that medical anthropology retrospectively recognised as its own. Numerous anthropologists (M. Augé, B. Hours, S. Fainzang, F. Laplantine, M. Perrin) devoted some attention to the study of illness, particularly its representations, without however always making it to their focus of interest. The philosopher Charles Canguilhem often provided a theoretical framework for their works. However, it was especially in psychiatry, during the mid-1960s, that a systematic encounter between the clinic and the anthropological approach began (Collomb and Zempléni 1965; Zempléni and Rabain 1965). The aim was to better understand the psychiatric clinic in an African context. This research was undertaken in the hospital of Fann (Dakar, Sénégal), at the same time that the journal *Psychopathologie Africaine* was founded. It published a variety of theoretical approaches at the time on the theme of possession, which later became core to medical anthropology (in Ethiopia M. Leiris; in Haiti A. Métraux).

Apart from the North American academy, George Devereux's extensive teaching in Paris left a mark, not least by contributing to the interest of psychiatrists for anthropology. The close relationship with the University of Montréal, which, in collaboration with McGill University, led in the 1960s to the establishment of a programme in transcultural psychiatry (GIRAME, Groupe interuniversitaire de recherche en anthropologie médicale et ethnopsychiatrie), also had a tangible influence in France. It was the care for immigrants from sub-Saharan Africa that eventually led to the creation, on French territory, of counselling in which psychiatrists, psychologists and anthropologists collaborated. Tropical medicine was another realm of encounter between anthropology and medicine, particularly, within the field of parasitology. The epidemiology of parasitic infections depends both on ecological factors and socio-cultural practices. Socio-cultural epidemiology as a field ensued from the 1970s onwards.

The two currents (medical and anthropological) had a formal encounter in the early 1980s. With few exceptions, it was the physicians and psychologists with a degree in anthropology who created a 'medical anthropology' (A. Retel-Laurentin, A. Zempleni, J. Benoist, A. Epelboin, F. Meyer, D. Fassin, A. Hubert). However, there was resistance from the anthropologists against the term 'medical anthropology' as they anticipated an overly biomedical orientation in the formulation of research questions. They endorsed instead an 'anthropologie de la maladie' or 'anthropologie de la santé', and vestiges of their reluctance have remained documented to the present day. It was after the round table discussion of 1980 at the CNRS, 'Une anthropologie médicale en France?' (published in 1983), that the term 'medical anthropology' started to spread. In 1982 Epelboin founded the journal 'Bulletin d'ethnomédecine', open to all theoretical approaches in the field; it soon had both currents specify their position. A volume edited by Augé and Herzlich (1983) underlined how important representations of illness were for anthropological inquiry and clearly positioned their anthropological studies within the humanities, emphasising collaboration with sociologists and historians. A further workshop, which had a unifying effect, was organised in 1983 at the CNRS in Paris: 'Le premier colloque national pour l'anthropologie médicale', which led to two further major publications in 1987.

Many anthropologists were, for a long time, reluctant to any form of 'applied anthropology', resisting the idea of having biomedicine define the object of anthropological study. The AIDS epidemic changed the situation: the questions asked by anthropologists, and the important financial means that the epidemic made available to them, led many to actively partake in its research. The national association AMADES (Anthropologie médicale appliquée au développement et à la santé), which has organised to the present day biennial conferences in France or other countries, was founded in 1990 at the university of Aix-Marseille III, where, since the 1980s, applied medical anthropology had been systematically developed to the doctoral level. Nowadays, medical anthropology considers itself relevant both for foundational research in anthropology and for applied research. It is taught in Paris, Lyon, Bordeaux, Strasbourg, Brest and elsewhere, in departments of anthropology and, since the 1990s, also in faculties of medicine.

References to the vignette

Augé M., and C. Herzlich, eds. 1983. *Le sens du mal, anthropologie, histoire, sociologie de la maladie*. Paris: Archives contemporaines.

Benoist, J., ed. 2002. *Soigner au pluriel. Essais sur le pluralisme médical*. Paris: Karthala.

Canguilhem, C. 1966. *Le normal et le pathologique*. Paris: PUF.

Collectif. 1983. *Une anthropologie médicale en France?* Paris: CNRS.

Collomb, H., and A. Zempléni. 1965. Maladie mentale et acculturation. *Médecin d'Afrique Noire*, 22, no. 8: 293–6.

Collomb, H. 1966. Psychiatrie et cultures. *Psychopathologie Africaine* 2, no. 2: 259–94.

Leiris, M. 1958. *La possession et ses aspects théatraux chez les Ethiopiens de Gondar*. Paris: Plon.

Leriche, R. 1949. *La chirurgie, discipline de la connaissance*. Paris: La Diane française.

Lévi-Strauss, C. 1949. Le sorcier et sa magie. *Les temps modernes*. 41: 3–24

Lévi-Strauss, C. 1949. L'efficacité symbolique. *Revue de l'histoire des religions* 135, no. 1: 3–27.

Métraux, A. 1958. *Le vaudou haïtien*. Paris: Gallimard.

Nicolle, C. 1939. *Destin des maladies infectieuses*. Paris: PUF.

Retel-Laurentin, A. 1987. *Etiologie et perception de la maladie dans les sociétés modernes et traditionnelles*. Paris: L'Harmattan.

Zempléni, A., ed. 1987. Causes, agents, origines de la maladie. *L'Ethnographie* (Special Issue) 81: 96–7.

Zempléni, A., and J. Rabain. 1965. L'enfant *nit ku bon*. Un tableau psychopathologique traditionnel chez les Wolof et les Lebou du Sénégal. *Psychopathologie africaine* 1, no. 3: 329–441.

Jean Benoist, translated from French

http://dx.doi.org/10.1080/13648470.2012.688349

What about *Ethnomedizin*? Reflections on the early days of medical anthropologies in German-speaking countries

Ruth Kutalek[a], Verena C. Münzenmeier[b] and Armin Prinz[a]

[a]Medical University of Vienna, Vienna, Austria; [b]Formerly University of Zurich, Zurich, Switzerland

This paper gives an overview of the historical and theoretical developments of *Ethnomedizin* in Germany, Austria and Switzerland. It demonstrates the richness of this interdisciplinary field at the interface of anthropology, medicine and public health, and provides the unique perspectives of some of its founders from the 1970s onwards.

Introduction

In German-speaking countries the discourses in anthropology and medicine, which started in the 1950s, were subsumed under the label of 'Ethnomedizin', an all-encompassing term (Schröder 1978) that did not have the specific connotations of the English 'ethnomedicine' (e.g. Nichter 1992) but includes most of what is currently known as medical anthropology. Today, the terms 'Medizinethnologie' (see Rudnitzki, Schiefenhövel and Schröder 1977, 4) and 'Medizinanthropologie' (Prinz 1984; Sterly 1992) are used; some also recur to the Anglophone term 'Medical Anthropology' (see vignette by Dilger). German-speaking researchers have faced enormous difficulties at establishing medical anthropology at university level. Very similar traditions – firmly grounded in medicine, which address its social scientific and humanist aspects – such as history of medicine, medical ethics, social medicine, tropical medicine and medical sociology were already well-established (Hauschild 2010, 433), competing with medical anthropology as a scientific discipline.

Predating the above developments, the works of two influential medical doctors deserve to be mentioned: Rudolf Virchow (1821–1902) and Viktor von Weizsäcker (1886–1957). Especially in Anglophone public health, Virchow is remembered as a *Leitfigur*. Of modest origin, he was a social reformer and active politician for decades and dealt with the social, political and economic determinants of disease. 'Should medicine indeed fulfil her noble task', he repeatedly stressed, 'she must intervene in the great [realm of] political and social life' (Virchow 1856, 56). Von Weizsäcker, who became one of the main founders of psychosomatic medicine in Germany, had by the 1920s conceptualised an important theoretical framework that differentiated between disease and illness: 'We understand disease but we understand neither the distress of disease nor what is essential to the ill person' (Weizsäcker 1987 [1926], 16, translation RK)[1]. In his view, suffering was an existential threat and an elemental characteristic of being a human.

In the early 1950s, the Viennese medical doctor Erich Drobec (1919–2004), who had also studied ethnology, physical anthropology, prehistory and botany, coined the term *Ethnomedizin*: 'Ethnomedizin has to deal primarily with the rationale of primitive peoples' medicine, with their physiologic and medical knowledge and the resulting measures applied in different stages of the life cycle (pregnancy, birth, puberty etc). A special branch is the natives' use of remedies (medicinal plants etc.)' (Drobec 1955, 950, translation RK). Drobec criticized the over-emphasis of magic as disease-aetiology and the neglect of considering peoples' rationale of disease aetiology underlying their ways of healing (Drobec 1953). In particular, he criticized the physician and historian Erwin Ackerknecht, who considered the medicine of indigenous people primarily magico-religious and who specifically denied the capacity for empirical thought and perception of causalities among people practising indigenous medicine (e.g. Ackerknecht 1946, 468). 'One will have to approve the fact that primitives arrived at their therapeutic methods and remedies by pure rational means' (Drobec 1956, 202, translation RK). Drobec was interested in psychiatry and psychotherapy and positioned himself clearly against the clerical and nationalist *Wiener Kulturhistorische Schule* of Wilhelm Schmidt and Wilhelm Koppers, a fact that brought his career to a sudden end (Kutalek 2009b). Drobec was the first in attempting to institutionalize *Ethnomedizin* at a German-speaking university, but when his efforts were hindered he pursued a career in medicine.

The Non-institutionalisation of *Ethnomedizin* in Western Germany[2]

Joachim Sterly (1926–2001), an ethnologist and philosopher, in 1969 founded the *Arbeitsstelle für Ethnomedizin* (AfE) in Hamburg.[3] In the same year, he established a section in the German Anthropological Association (GAA) in Göttingen called *Arbeitsgruppe Ethnomedizin, Ethnobotanik und Ethnozoologie in der DGV (Deutsche Gesellschaft für Völkerkunde)*, which included important members who held chairs, like Lorenz Löffler in Zurich, Otto Zerries in Munich, Sigrid Paul in Salzburg, Erhard Schlesier in Göttingen, and others who introduced new interdisciplinary perspectives on health and illness that had not at all been common to the academic communities of the *Völkerkunde* of those days. Sterly's initiative, which drew on Drobec's term and which introduced Husserl and his concept of *Lebenswelt* into the academic debates it ignited (Sterly 1974), was quickly and well received by physicians interested in the social aspects of their work, and by German anthropologists who were interested in the effects of decolonization and rapid social change in urban communities (Luig 1978), and in the question of growing populations (Löffler 1979).

In 1970, Sterly also founded the *Arbeitsgemeinschaft Ethnomedizin* (AgE, later AGEM) and in 1971 the journal *Ethnomedizin: Zeitschrift für interdisziplinäre Forschung* (Sterly 1971). The latter had an internationally renowned advisory board, recruited its authors worldwide, and existed until 1982. Apart from Sterly there were two other managing editors: Gerhard Rudnitzki, a social psychiatrist from Heidelberg, and Werner Stöcklin, a paediatrician from Basel who had worked in Papua New Guinea for a long time (e.g., Stöcklin 1984). AGEM was founded in order to unify anthropologists, physicians, psychologists and psychiatrists whose research emphasized the interface of culture and medicine. The first conference of the group was held in Munich in 1973, the following two in Heidelberg, in 1974 (Schröder 1977) and 1977. *Curare*, with the subtitle *Ethnomedizin und Transkulturelle Psychiatrie*, became the new journal of the AGEM (www.agem-ethnomedizin.de). It was founded in 1978 by Gerhard Rudnitzki, along with Beatrix Pfleiderer, Wulf Schiefenhövel and Ekkehard Schröder (the latter two also medically trained).

The South Asian Institute (SAI), founded in 1962 in Heidelberg, encompassed all academic branches researching South Asia, including cultural anthropology, tropical medicine and public health. Heidelberg thus became known for carrying out research within this frame from the early 1970s. The first German graduates of public health returning from London, Liverpool or Harvard brought with them new notions of communal medicine and public health. In 1973, Hans-Jochen Diesfeld, who was chair of Tropical Hygiene and Public Health at Heidelberg from 1975 to 1997, initiated an important international conference in Berlin (Diesfeld and Kroeger 1974). Having learned about basic health needs as medical officer in the Haile-Selassie Hospital in Addis Ababa in the 1960s, he developed new ways of thinking in medicine, integrating ecological, epidemiological, socio-economic and cultural perspectives to define the essential fields of medical care. His work laid the foundations for a new course about medicine in developing countries, which granted the dynamics of poverty a defined place in the field, an approach that was controversial and revolutionary in the medical field of those days (Bichmann 1992).

In 1974, Diesfeld also initiated a course on *Medizin in Entwicklungsländern*, held twice a year during four-week periods, which was geared towards medical graduates assigned to work in so-called 'developing countries'; it was supported by all the important German government- and non-government organisations in the health sector. Diesfeld's assistant lecturer Ekkehard Schröder, who had studied medicine and anthropology in Kiel and Heidelberg, convened the courses for three periods (1976–1978) and had the task of designing the cultural dimension of the curriculum (Diesfeld and Schröder 1978). In winter 1977/78, Schröder gave the first lectures in *Ethnomedizin* at a German university, and in that year the term *Ethnomedizin* entered into the famous medicinal dictionary *Pschyrembel*; it also became very popular through Ludwig and Pfleiderer-Becker's (1978) widely used introductory reader *Materialien zur Ethnomedizin*, which had a representative international bibliography, and, in particular, through Pfleiderer and Bichmann's (1985) *Krankheit und Kultur*.

Diesfeld granted an academic position also to Dorothea Sich, who from 1964–1976 during her clinical work as a gynaecologist in Korea had already realized the importance of cultural dimensions in doctor–patient interactions (e.g., Sich 1979). She developed a complete set of teaching modules called '*Kulturvergleichende Medizinische Anthropologie*' (Sich et al. 1994). She was partly assisted by cultural anthropologist Paul Hinderling (1981), a collaborator of Ernst E. Boesch (1972); Boesch had a chair in psychology in Saarbrücken and founded there the *Sozialpsychologische Forschungsstelle für Entwicklungsplanung*. In the 1980s nearly all German researchers of the older generation stopped at Heidelberg, either holding a teaching or research position for some time or completing a PhD or MD thesis (e.g. Beatrix Pfleiderer, Wolfgang Bichmann, Katarina Greifeld, Thomas Lux, Johannes Sommerfeld), and the current professor of anthropology, William Sax, builds on this legacy (Sax 2011). Regular seminars in *Ethnomedizin* were also held in Freiburg by the anthropologist and psychiatrist Winfried Effelsberg (2005) and the botanist and anthropologist Michael Heinrich, and in Munich by the anthropologist and physician Wulf Schiefenhövel. In 1982, Beatrix Pfleiderer, who had done extensive fieldwork in India since the 1970s, obtained a chair at the University of Hamburg where she then taught 'Medizinethnologie' until 1992 (when she embarked on a new career in Hawaii). She organised an important international conference on 'Anthropologies of Medicine' (Pfleiderer and Bibeau 1991).

Ethnomedizin in Austria

When, inspired by Erich Drobec, Armin Prinz became interested in *Ethnomedizin*, it had not been a research topic for almost two decades. In the 1970s, he and Manfred Kremser did their first fieldwork on nutritional anthropology (Prinz 1976) and witchcraft (Kremser 1977) among the Azande in the northeast of what is now the Democratic Republic of Congo (former Zaire). This brought them into contact with Sir Edward Evans-Pritchard.[4] In 1978, Prinz, a medical doctor and social anthropologist, founded the *Oesterreichische Ethnomedizinische Gesellschaft*, which emphasized interdisciplinarity among scholars from the three fields of medicine, anthropology, and the natural sciences (Prinz 2009). In 1981, he started giving the first introductory lectures in *Ethnomedizin* in Vienna. He became Professor in 1989, being one of only six persons to habilitate (*venia legendi*) in *Ethnomedizin* during those years. In 1993, the *Abteilung Ethnomedizin* was founded in the Institute of the History of Medicine in the Faculty of Medicine; it was implemented by Erhard Busek, at that time Minister of Science. The Institute of the History of Medicine, when headed by Erna Lesky (1911–1986), contributed the first monographs towards the systematic building up, since 1988, of a specialist library listing more than 11,000 titles. Today it is the largest library within the field in the German-speaking countries. The *Viennese Ethnomedicine Newsletter* and two monograph series followed suit in 1996. The Ethnomedical Collection, which consists of more than 1800 objects related to ethnomedicine and focuses on paintings from internationally known African artists, was founded in 1999. Since 2008 the now-called 'Unit Ethnomedicine and International Health' has been part of the Department of General Practice and Family Medicine, Center for Public Health, Medical University of Vienna. Teaching activities have always attracted numerous students from social anthropology, medicine and elsewhere. Since the 1980s, electives in ethnomedicine, nutritional anthropology and ethnopharmacology have been offered on a regular basis, first at the *Institut für Voelkerkunde*, then at the Institute of the History of Medicine, and more recently, *Ethnomedizin/Medical Anthropology* has become a mandatory subject in the curriculum of undergraduate medical students, including topics such as global health, health and migration, death and dying, and gender (Kutalek 2009a, 2011). Apart from the 'Unit Ethnomedicine', *Ethnomedizin/Medical Anthropology* has also been taught at the Department of Social and Cultural Anthropology, University of Vienna, Austria's only anthropology department, on a larger scale since 2002 when the teaching module MAKOTRA (*Medical Anthropology, body awareness, transculturality*) gave the subject a prominent place in both undergraduate and post-graduate training. Research in *Ethnomedizin/Medical Anthropology* emphasizes regional health systems, concepts of disease and medical pluralism (Kremser 1977; Prinz 1976, 1986; Sabernig 2007; Schaffler 2009), applied medical anthropology (Burtscher 2009), women's health (Binder-Fritz 1996), new reproductive technologies and other biomedical practices (see vignette by Hadolt).

Initiating *Ethnomedizin* in Switzerland in the 1970s and 1980s

In contrast to Austria and Germany, courses in *Ethnomedizin* were first delivered within social anthropology in Switzerland, at the *Ethnologisches Seminar* of the University of Zurich.[5] Up until the 1970s, cultural/social anthropology, or *Völkerkunde* as it was then called, had been strongly associated with the ethnographic museum, a position that defined its emphasis on material culture and its main focus on small-scale societies. In 1971 the University of Zurich instigated a separation between social/cultural anthropology and geography in order to create adequate training and, thus, better working opportunities for the increasing number of students (Universität Zürich, *Strukturbericht* 24 June 1970). An independent, practice-oriented department was established, which had as its main focus culture change and work in developing countries. Lorenz G. Löffler, who had been professor at the South Asia Institute in Heidelberg (Germany), became the chair of this new department. Under his guidance the new institutional structure developed rapidly in a favourable social climate. Following the prolonged social unrest of the youth movement of the 1960s, the students of the 1970s and 1980s had new expectations, and a keen interest in the social sciences. Hence, the number of students increased from 75 to 670 in only ten years, and the courses from 7 to 31 in 13 years (Universität Zürich, lecture lists of 1971 and 1984). The attendance was tremendous; often big lecture halls had to be used for the more than 300 students and – in the case of the completely new course in urban anthropology – even the largest auditorium was not big enough for the 650 attendants. To guarantee quality training, Löffler carefully chose his assistants and temporary lecturers, and he gave them great responsibility and unparalleled freedom in developing new study and research programmes. *Ethnomedizin* had been introduced early as an option course alongside kinship, ethnomusicology, economics, politics, ethnopsychology and psychoanalysis, customary law, conflict, gender relations and the like. Furthermore, undergraduates were encouraged to undertake a year of field research, as training in methodology, in a so-called developing country. This feature produced new geographic focuses. In this way, Zurich became the leading department in Switzerland, offering the largest range of courses in what was called *Ethnologie*.

As his first assistant lecturer in *Ethnomedizin*, Löffler employed Verena Kücholl (who now calls herself Verena C. Münzenmeier); she had been his student and had professional medical experience. She had conducted one year of fieldwork (1974/75) in Bangladesh, where she had studied the living conditions of the Oraõn and the life of women – several of them active in basic health care. Immediately after her graduation in 1977, she started lecturing on this new sub-field of social anthropology. Until finishing her doctorate in 1984 (Kücholl 1985a), she taught four cycles of two or three semesters and started networking with, for instance, the AGEM in Germany and its journal *Curare*. The bibliography was selected according to its usefulness for students doing fieldwork on medical systems. Apart from publications in German, mainly Anglo- and Francophone materials were used. The work of

Erwin Ackerknecht (1946) was already an asset for Zurich, but as *Ethnomedizin* was becoming fashionable, new introductions and readers as well as case studies were available. One part of the lectures concentrated on traditional healing and topics such as shamanism, magic, witchcraft, trance, divination, ritual healing, training, and indigenous practices of birth; another part concerned western medicine and the development of national health systems at central and local levels, the training of medical staff and village health workers, and public health issues such as epidemics, poverty, malnourishment, water supply and hygiene. The courses contained a wealth of comparative materials, and as Kücholl was particularly interested in ritual healing practices, this topic became the heart of her own research (Kücholl 1985a: see in particular 34, 143).

In the 1980s, Swiss social anthropologists were generally met with scepticism; their work was little known and often considered a hobby. In order to investigate whether western medicine had superseded traditional medicine after 60 years of cooperation in the African highland Kingdom of Lesotho, and which treatment the different population groups preferred, the Swiss association 'Friends of Lesotho' in 1984/85 asked Kücholl to carry out an 18-month study (Kücholl 1985b). Although proposed by a Swiss physician, her research was initially met with stiff opposition. The doctors were convinced that non-doctors could not assess their work, and that they themselves would best know its effects. The study showed that wealthier people of this basically poor population benefited from both medical systems, while the poorest were excluded from both. Traditional medicine was the priority. It became clear that bride price payments with cattle were linked to the healing process, in that it led to the stabilization of a marriage and of the relations between the families, and thus served as a shield against impoverishment The ancestors, embodied in cattle, supervised these payments over time and protected their descendants: healing and social security in one (see also Löffler 1979). This tradition stood in sharp contrast to western views that women should not be 'sold' and that it was superstitious to believe in the healing qualities of ancestors. For the local health personnel, the study provided intellectual support for explaining the beneficial effects of traditional healing. On the other hand, Kücholl's findings about medical needs, the use of medical facilities and questions of compliance, helped in the revision and implementation of the Primary Health Care strategy, and laid the basis for her active involvement as an applied medical anthropologist in development co-operations until the end of her professional career.

It is Löffler's merit to have promoted medical anthropology for more than 20 years until his retirement in 1995. His appointments started with Kücholl and continued with her former students and their research specialties. Thus, Sherill Freeman followed from 1984–86 concentrating on psychoanalytic aspects, and Martine Verwey was repeatedly engaged as temporary lecturer from 1987–95; her special research topic was the contribution of medical anthropology to migration research (Verwey 2001). In high demand were also Liselotte Kuntner's lectures on childbirth and maternity in cross-cultural perspective from 1989–99 (Kuntner 1985). When Elisabeth Hsu, after completion of her doctorate (Hsu 1999 [1992]), became assistant lecturer from 1992–95, Löffler aimed to institutionalize the courses more firmly in the curriculum but after his retirement this plan was no longer pursued. By that time anthropologists at other Swiss universities had contributed substantially to the quick spreading of the field (e.g. Gonseth 1993/1994). Importantly, Brigit Obrist

(Basel), Martine Verwey (Zurich) and Hans-Rudolf Wicker (Berne), and others, founded an inter-university commission in 1992 (see vignette by van Eeuwijk).

Acknowledgements

The authors appreciate the extensive editorial work that has gone into this paper. The Austrian part of the paper was presented at the conference 'Medical anthropology in Europe' funded by the Wellcome Trust and Royal Anthropological Institute.

Conflict of interest: none.

Notes

1. His book *Der Gestaltkreis* was translated into French by Michel Foucault and Daniel Rocher (Weizsäcker 1958 [1940]). Translations of part of his work exist, e.g. in Japanese, Spanish and Italian, but not in English.
2. This section is based on the writing and information provided by Ekkehard Schröder.
3. The first three newsletters of AfE were reprinted in 2010 issues of *Curare* 33, no. 1+2: 135–52. On early developments, see also Schröder (2002).
4. In autumn of 1973, Evans-Pritchard had planned to come to Vienna but unfortunately died on 11 September of the same year (letter to Armin Prinz).
5. For more detail, see Münzenmeier (2011).

References

Ackerknecht, E.H. 1946. Natural diseases and rational treatment in primitive medicine. *Bulletin of the History of Medicine* 19: 467–97.

Bichmann, Wolfgang, ed. 1992. *Querbezüge und Bedeutung der Ethnomedizin in einem holistischen Gesundheitsverständnis. Festschrift zum 60. Geburtstag von Hans-Jochen Diesfeld*. (Special issue). *Curare* 15, no. 1+2: 1–144.

Binder-Fritz, Christine. 1996. *Whaka Whanau. Geburt und Mutterschaft bei den Maori in Neuseeland*. Bern: Lang.

Boesch, Ernst E. 1972. *Communication between doctors and patients. Part I: Survey of the problems and analysis of the consultations (Thailand)*. Saarbrücken: Sozialpsychologische Forschungsstelle für Entwicklungsplanung.

Burtscher, Doris. 2009. Men – Taking risk or taking responsibilitiy. Knowledge and perceptions of HIV/AIDS. The Ndebele in the Tsholotsho district, Zimbabwe. In *Essays in medical anthropology: The Austrian Ethnomedical Society*, ed. Ruth Kutalek and Armin Prinz, 249–67. Münster: Lit.

Diesfeld, Hans-Jochen, and Erich Kroeger, eds. 1974. *Community health and health motivation in South East Asia* (Beiträge zur Südasienforschung 4). Wiesbaden: Steiner.

Diesfeld, Hans-Jochen, and Ekkehard Schröder, eds. 1978. *Medizin in Entwicklungsländern. Ein praxisorientierter Vorbereitungskurs für Ärzte, die erstmalig in Entwicklungsländern tätig werden*. Ergänzende Skripten (Ringbuch, 1. Auflage). Heidelberg: Institut für Tropenhygiene.

Drobec, E. 1953. Beiträge zur Ethnomedizin. *Wiener Völkerkundliche Mitteilungen* 1, no. 2: 57–60.

Drobec, E. 1955. Zur Geschichte der Ethnomedizin. *Anthropos* 50: 950–7. Reprinted in 2005 in *Curare* 28, no. 1: 3–10.

Drobec, Erich. 1956. Beiträge zur Methode der Ethnomedizin. In *Festschrift anläßlich des 25-jährigen Bestandes des Institutes für Völkerkunde der Universität Wien (1929–1954)*, eds. J. Haekel, A. Hohenwart-Gerlachstein, and A. Slawik, 193–204. Wien: Ferdinand Berger.

Effelsberg, W. 2005. Ist die Ethnomedizin eine Interdisziplin? Gedanken zu alten und neuen Studiengängen und zur interkulturellen Kompetenz im Fache Soziale Arbeit. *Curare* 28, no. 1: 96–8.

Gonseth, Marc-Olivier, ed. 1993/1994. *Les frontières du mal: approaches anthropologiques de la santé et de la maladie/Kranksein und Gesundwerden im Spannungsfeld der Kulturen.* Bern: Lang. (Ethnologica Helvetica, 17–18).

Hauschild, Thomas. 2010. Ethnomedizin, medizinische Ethnologie, Medizinanthropologie: Erfolge, Misserfolge und Grenzen. In *Medizin im Kontext*, eds. H. Dilger and B. Hadolt, 431–439. Frankfurt/Main: Lang.

Hinderling, Paul. 1981. *Kranksein in 'primitiven' und traditionalen Kulturen.* Norderstedt: Verlag für Ethnologie.

Hsu, Elisabeth. 1999 [1992]. *The transmission of Chinese medicine.* Cambridge: Cambridge University Press.

Kremser, Manfred. 1977. Hexerei (mangu) bei den Azande. Ein Beitrag zum Verständnis kulturspezifischer Krankheitskonzeptionen eines zentralafrikanischen Volkes. Dissertation Universität Wien.

Kuntner, Liselotte. 1985. *Die Gebärhaltung der Frau: Schwangerschaft und Geburt aus geschichtlicher, völkerkundlicher und medizinischer Sicht.* München: Hans Marseille (1994 enlarged 4th edn).

Kutalek, Ruth. 2009a. Medical anthropology in medical education – a challenge. In *Essays in medical anthropology: The Austrian Ethnomedical Society*, ed. Ruth Kutalek and Armin Prinz. Vienna: Lit.

Kutalek, R. 2009b. Erich Drobec – Pionier der Ethnomedizin im Spannungsfeld Religion und Medizin. *Anthropos* 104: 527–33.

Kutalek, Ruth. 2011. *Medizinanthropologie am Schnittpunkt. Interdisziplinarität zwischen Anthropologie und Medizin.* Habilitationsschrift, Universität Wien.

Kücholl, Verena. 1985a. *Soziokulturelle Wege des Heilens: eine ethnomedizinische Analyse und Interpretation des Samkhya und der Heiltradition der Navajo.* (Medizin in Entwicklungsländern 20). Frankfurt/Main: Lang.

Kücholl, Verena. 1985b. *Ethnomedical evaluation in Lesotho, 1984–1985. Final report submitted to the Private Health Association of Lesotho (PHAL) and the Swiss Association Friends of Lesotho.* Thaba-Tseka, Lesotho: Paray Hospital.

Löffler, L.G. 1979. Bevölkerungswachstum und Systeme sozialer Sicherung. *Curare* 2: 141–62, Reprinted in 2002 in *Curare* 25, no. 1+2: 243–59 and in Löffler, Lorenz G. 2002 *Aussaaten. Ethnologische Schriften.* Zürich: Argonaut.

Ludwig, Bruni, and Beatrix Pfleiderer-Becker, eds. 1978. *Materialien zur Ethnomedizin.* Bensheim: Kübel-Stiftung (1981, 2nd edn).

Luig, U. 1978. Sorcery accusations as social commentary. A case study of Mulago/Uganda. *Curare* 1, no. 1: 31–42, reprinted in 2008 in *Curare* 31, no. 2+3: 231–9.

Münzenmeier, V.C. 2011. Rasante Entwicklung der Ethnologie und Ethnomedizin in Zürich in den 70er und 80er Jahren. *Curare* 34, no. 3: 237–42.

Nichter, Mark, ed. 1992. Ethnomedicine: Diverse trends, common linkages. *Anthropological approaches to the study of ethnomedicine.* Yverdon: Gordon and Breach.

Pfleiderer, Beatrix, and Bichmann Wolfgang. 1985. *Krankheit und Kultur. Eine Einführung in die Ethnomedizin.* Berlin: Reimer.

Pfleiderer, Beatrix, and Gilles Bibeau, eds. 1991. *Anthropologies of medicine. A colloquium on Western European and North American perspectives.* (*Curare* Special Volume 7). Braunschweig: Vieweg.

Prinz, Armin. 1976. Das Ernährungswesen bei den Azande Nordost-Zaires. Ein Beitrag zum Problem des Bevölkerungsrückganges auf der Nil-Kongo-Wasserscheide. Dissertation, Universität Wien.

Prinz, A. 1984. Die Ethnomedizin. Definition und Abgrenzung eines interdisziplinären Konzeptes. *Mitteilungen der Anthropologischen Gesellschaft in Wien* 114: 37–50. Reprint in 1992 in *Curare* 15:147–60.

Prinz, Armin. 1986. Initialerlebnis und Heilberufung. In *Traditionelle Heilkundige – Ärztliche Persönlichkeiten im Vergleich der Kulturen und medizinischen Systeme* (*Curare* Special Volume 5). Braunschweig: Vieweg.

Prinz, Armin. 2009 The history of the Austrian Ethnomedical Society (Österreichische Ethnomedizinische Gesellschaft). In *Essays in medical anthropology: The Austrian Ethnomedical Society*, ed. Ruth Kutalek and Armin Prinz, 3–13. Vienna: Lit.

Rudnitzki, Gerhard, Wulf Schiefenhövel, and Ekkehard Schröder, eds. 1977. *Ethnomedizin – Beiträge zu einem Dialog zwischen Heilkunst und Völkerkunde.* Barmstedt: Detlev Kurth.

Sabernig, Katharina. 2007 *Kalte Kräuter und heiße Bäder. Die Anwendung der Tibetischen Medizin in den Klöstern Amdos.* (Wiener ethnomedizinische Reihe 5). Vienna: Lit.

Schaffler, Yvonne. 2009. *Vodú? Das ist Sache der anderen! Kreolische Medizin, Spiritualität und Identität im Südwesten der Dominikanischen Republik.* (Wiener ethnomedizinische Reihe 7). Vienna: Lit.

Sax, W. 2011. Medical anthropology at Heidelberg. *Viennese Ethnomedicine Newsletter* XII (2–3): 3–5.

Schröder, Ekkehard, ed. 1977. *Faktoren des Gesundwerdens in Gruppen und Ethnien.* Verhandlungen des 2. Rundgespräches 'Ethnomedizin' in Heidelberg vom 29. bis 30. 11. (Beiträge zur Südasienforschung 30). Wiesbaden: Steiner.

Schröder, E. 1978. Ethnomedicine and medical anthropology. A survey of developments in Germany (from a viewpoint in 1978). *Reviews in Anthropology* 5, no. 4: 478–85. Reprinted in *Essays in medical anthropology: The Austrian Ethnomedical Society*, ed. R. Kutalek and A. Prinz, 55–68. Münster: Lit.

Schröder, Ekkehard, ed. 2002. *Der frühe ethnomedizinische Diskurs in der Curare. Ausgewählte Artikel aus den ersten 12 Jahren der Zeitschrift Curare, gewidmet dem Gründer der AGEM, Joachim Sterly (1926–2001)* (Special issue) *Curare* 25, no. 1+2.

Sich, D. 1979. Naeng. Begegnung mit einer Volkskrankheit in der modernen frauenärztlichen Sprechstunde in Korea. *Curare* 2, no. 2: 87–96. Reprint in 2002 in *Curare* 25, no. 1+2: 121–8.

Sich, Dorothea, Hans-Jochen Diesfeld, Angelika Deigner, and Monika Habermann, eds. 1994. *Medizin und Kultur. Eine Propädeutik für Studierende der Medizin und der Ethnologie mit 4 Seminaren in Kulturvergleichender Medizinischer Anthropologie (KMA).* (Medizin in Entwicklungsländern 34). Frankfurt/Main: Lang.

Sterly, J. 1971. Ethnomedizin. Entwurf einer Zeitschrift. *Ethnomedizin* 1, no. 1: 7–9. Reprint in 2001 in *Curare* 24, no. 1+2: 230.

Sterly, J. 1974. Zur Wissenschaftstheorie der Ethnomedizin. *Anthropos* 69: 608–15.

Sterly, J. 1992. Das Anliegen der deutschsprachigen Medizinanthropologie in den Ländern der Europäischen Gemeinschaft. *Gesundheitswesen* 54: 341–5.

Stöcklin, Werner. 1984. *Toktok: am Rande der Steinzeit auf Neuguinea.* Basel: Birkhäuser. (2004 3rd expanded edition).

Verwey, Martine, ed. 2001. *Trauma und Ressourcen/Trauma and Empowerment.* (*Curare* Special Volume 16). Berlin: Vieweg.

Virchow, Rudolf. 1856. *Gesammelte Abhandlungen zur wissenschaftlichen Medicin.* Frankfurt am Main: Verlag von Meidinger Sohn & Comp.

Weizsäcker, Viktor von. 1987 [1926 etc.]. *Gesammelte Schriften. Der Arzt und der Kranke. Stücke einer medizinischen Anthropologie.* Vol. 5, eds. Peter Achilles, Dieter Janz, Martin Schrenk, and Carl Friedrich von Weizsäcker. Frankfurt/Main: Suhrkamp.

Weizsäcker, Viktor von. 1958 [1940]. *Le cycle de la structure. (Der Gestaltkreis),* translated by M. Foucault and D. Rocher. Bruges: Desclée de Brouwer (2nd edn 1999).

Vignettes

Medical anthropology: the working group within the German Anthropological Association

The work group 'Medical Anthropology' (www.medicalanthropology.de) was established as a subsection of the German Anthropological Association (DGV) in 1997, with members mainly from Germany, Switzerland and Austria. In the beginning, it was primarily an initiative of master's and doctoral students who felt the need to create a forum for disciplinary as well as interdisciplinary exchanges. Many of the members had conducted field research on medical anthropological topics, but were attached to academic environments with only a marginal interest in the sub-discipline. Networking was a first goal of the newly established group, as was the development of ensuing thematic workshops and conferences. The systematic integration of medical anthropological perspectives into teaching and research at German-speaking universities has since become of foremost importance. The choice of the Anglophone term 'Medical Anthropology' was deliberate: *Ethnomedizin* at the time had been shaped by biomedical agendas and focused mostly on the study of medical beliefs and health systems in non-western countries; *Medical Anthropology*, on the other hand, referred to the established research tradition in the Anglophone world with its firm basis in the theoretical and methodological foundations of social and cultural anthropology. Over the years, the work group has grown rapidly and today comprises more than 90 members in all three German-speaking countries. Between 1997 and 2011, it held 15 national and international workshops and conference panels. Key publications include Wolf and Hörbst (2003) and Dilger and Hadolt (2010). The group has now established numerous linkages with medical anthropologists in Europe and beyond, and it was formally involved in the steering group for establishing the EASA Network for Medical Anthropology that was called into life in 2006.

References

Dilger, Hansjörg, and Bernhard Hadolt, eds. 2010. *Medizin im Kontext: Krankheit und Gesundheit in einer vernetzten Welt*. Frankfurt/Main: Lang.
Wolf, Angelika and Viola Hörbst, eds. 2003. *Medizin und Globalisierung: universelle Ansprüche – lokale Antworten*. Münster: Lit.

Hansjörg Dilger, Freie Universität Berlin

http://dx.doi.org/10.1080/13648470.2012.688350

Medical anthropology in Switzerland

'Medical Anthropology Switzerland' (MAS) was founded in 1992 (www.medicalanthropology.ch) and is an active committee of the Swiss Ethnological Society. A few years earlier (1989), a small group of social anthropologists from various Swiss universities had begun to discuss a more intensified exchange and coordination of medical anthropology issues. They created a special interest group representing the country's different regional cultures (thus referring to a French, a German and an Italian history of medical anthropology), the different professional fields that deal in a very broad sense with 'culture, health and illness' (e.g. family practitioners, psychiatrists, social anthropologists, sociologists, and nurses), and the different academic stages such as taught students, PhD researchers and postdoctoral seniors. To emphasize that this group had an interdisciplinary orientation but saw medical anthropology to be firmly grounded in anthropology, it called itself 'Interdisziplinäre Kommission für Medizinethnologie' (IKME) – not 'Ethnomedizin' – and since 2002 'Medical Anthropology Switzerland' (MAS). The anthropological roots, interdisciplinary orientation and intercultural balance were documented in several readers (e.g. Moser, Nyffeler, and Verwey 2001; Eeuwijk and Obrist 2006), and they have been maintained up to now. Every year, MAS organizes an international symposium and at least two smaller workshops. The initial broad scope of interest is still oscillating between a university-based, theory-driven and epistemological medical anthropology and an applied medical anthropology.

References

Eeuwijk, Peter van, and Brigit Obrist, eds. 2006. *Vulnerabilität, Migration und Altern. Medizinethnologische Ansätze im Spannungsfeld von Theorie und Praxis*. Zürich: Seismo.

Moser, Catherine, Doris Nyffeler, and Martine Verwey, eds. 2001. *Traumatisierung von Flüchtlingen und Asyl Suchenden. Einfluss des politischen, sozialen und medizinischen Kontextes*. Zürich: Seismo.

Peter van Eeuwijk, Universities of Basel, Freiburg i. Br and Zürich

http://dx.doi.org/10.1080/13648470.2012.688351

Studying medical technologies

Biomedical technologies became one of medical anthropology's major topics only in the 1990s, when new medical technologies such as assisted reproductive technologies, organ transplantation and genetic testing emerged forcefully on a global scale. An example of extensive ethnographic research in Austria followed involuntarily childless couples in their quest for the own child over one and a half years, showing how peoples' views and actions change over time as they become experienced – and sometimes happy, but all too often frustrated – users. Through an overly pragmatic attitude, they often feel urged to 'try out' any promising technology before they can admit the hopelessness of their efforts, with the result that assisted reproductive technologies, by shaping the experiences, ideas and practices concerning bodies and parenthood, become a family- and partnership-building technology rather than simply a technology for conceiving children (Hadolt and Lengauer 2003). Utilizing Schatzki's 'site ontology', the same authors (Hadolt and Lengauer 2009) have mapped the landscape of pre-symptomatic genetic counselling and testing by relating their ideological and organizational specificities to their professional sitement. Whereas more traditional counselling modes sited in medical genetics aim at providing 'good counselling', other counselling modes sited in clinical departments, where genetics play a secondary role, aim at 'good treatment'.

References

Hadolt, Bernhard, and Monika Lengauer, 2003. Kinder-Machen: Eine ethnographische Untersuchung zur Handhabe von ungewollter Kinderlosigkeit und den Neuen Reproduktionstechnologien durch betroffene Frauen und Männer in Österreich. Dissertation, Universität Wien.

Hadolt, Bernhard, and Monika Lengauer, 2009. *Genetische Beratung in der Praxis: Herausforderungen bei präsymptomatischer Gendiagnostik am Beispiel Österreichs.* Frankfurt/Main: Campus.

Bernhard Hadolt, University of Vienna

http://dx.doi.org/10.1080/13648470.2012.688352

Part II

INTRODUCTION TO PART II

Medical anthropology in Europe – *quo vadis?*

Elisabeth Hsu

Institute of Social and Cultural Anthropology, University of Oxford, Oxford, UK

In the 1960s and 1970s, European Universities were affected, if not enduringly transformed, by anti-hegemonic social movements, such as the feminist movement and women's liberation; the anti-psychiatry movements; the Prague Spring and the Spring in Paris of 1968; the Club of Rome's dire ecological prognosis; and the anti-nuclear, environmental, alternative medical, anarchist and other leftist ideas, which spurned intense discussion and, sometimes, action. The economies in Europe were thriving at the time, but there was a sense that the consumer society they engendered would not be enduring; social crisis was imminent, if it had not already affected some economies.

Social/cultural anthropology gained new impetus in these climates of higher education seeking for other modes of knowledge production than the bureaucratic ones institutionalised at universities. The anthropological field method was not reductionist but required full-time and long-term immersion in research that would not leave anyone personally unaffected. Its results, which were later reduced to being termed 'qualitative', implicitly critiqued undue reliance on numbers and statistics. Finally, its basic assumption that through the study of another society, one would become more self-aware and critical of one's own society, drew many a student in search of a better world.

The current topics ranging from 'the practice of care' to 'the body politic' and 'the psy-dimension of personhood', which are discussed in the second part of this special issue, owe perhaps more than is generally acknowledged to these social movements. In Europe, 'the practice of care' has been studied from many angles, and importantly also from a perspective that aims to overcome the somewhat artificial division between 'applied' and more 'theoretically-oriented' research often referred to. The focus on practice that it implies is best understood in light of 'the problem of knowledge', which was central to early discussions in the field. The second theme, on the 'body politic', owes much to the feminist movement, which uncovered gendered asymmetries and also led to a re-evaluation of the body (as the female counterpart to the male mind) by stressing the importance of researching the body as a cultural and socio-political project in the making. Finally, the anti-psychiatry movements, which highlighted the irrationality of the concept 'mental disease', engendered innovative medical anthropological research into 'the psy-dimension of personhood'.

Medical anthropology emerged as an academic field at a time when these social movements shook Europe. From these days emerged a concept of holism that, in place of the logical opposite, monism, became the catchall to overcome problems caused by an overly rationalistic outlook for which the dualist Cartesian framework was made responsible. Medical anthropology attracted many a student interested in other, non-Western, Amerindian or Far Eastern, shamanic or scholarly ways of doing medicine. The social, and at times Marxist, critique of one's own society was an important incentive.

The problem of knowledge

'The problem of knowledge' arises when one appreciates magical and medical knowledge and practice for their, if only partial, internal coherence and validity. It has been addressed from successively different angles – including, very roughly speaking, the ethno-sciences in the 1960s, the anthropology of knowledge in the 1970s, the sociology of knowledge in the 1980s, and Science and Technology Studies (STS) since the 1990s.

Ethnoscientific endeavours, which were often undertaken by linguists and/or psychologists, engendered the concept of an 'ethnomedicine' alongside ethnobotany and ethnozoology (e.g. Fabrega 1975). However, already the earliest articles that were to define the field of medical anthropology expressed doubts about any decontextualised taxonomy of knowledge. In the first issue of *Culture, Medicine and Psychiatry*, Good (1977) criticised Frake's (1961) classification of Subanun skin diseases for ordering knowledge taxonomically. Instead, he explained the 'heart distress' of women participating in a family planning project in semantic networks, thereby aiming to analyse medical knowledge as constitutive of societal workings at large. Furthermore, an anthropological analysis was not to over-systematise, a legacy in Europe strongly associated with Barth (1975), which has remained an issue of undiminished concern (e.g. Littlewood 2007).

During the heyday of the rationality debate, semantic networks and explanatory models (EMs) came out of Harvard, while in the UK Horton and Finnegan's (1974) *Modes of thought*, Gellner's (1974) *Legitimation of belief* and Needham's (1975) 'Polythetic classification', which all built on Evans-Pritchard's legacy, inspired research on illness causation, nosological categories and other 'systems' of knowledge and practice. But as these debates started to become polarised into either 'universalist' or 'relativist' claims, researchers were alienated. Instead they increasingly turned to practice theory.

Furthermore, it would appear that the critique of modernity, which is perhaps implicit in most anthropological studies of another society, became increasingly more explicit. It was fuelled by an othering of 'The other' and a self-critical stance edging on the self-destructive. By the 1990s, it had effected a shift towards critiquing modern scientific medicine in Europe and North America. This development in thematic orientation coincided with an increasing uncertainty about those economies; field research abroad was no longer always encouraged. The shift in focus from the clinic in the tropics to the laboratory next door put centre stage the sociology, rather than the anthropology, of knowledge. It began with a social constructivist critique of science, highlighting the importance of practical tacit knowledge (Latour and Woolgar 1979), and deconstructed medicine as a monolithic enterprise (e.g. Berg and Mol 1998). Although neither of these insights were radically new, as evidenced by Polanyi (1958) and the authors in Part I of this volume, the framing of the inquiry was intriguing on both a methodological and theoretical level. It has since further accentuated the shift away from epistemologies to ontologies (e.g. Mol 2002) and spearheaded research into new socialities (e.g. Franklin 1997). Medical anthropologists from London to Vienna, Amsterdam to Edinburgh, and elsewhere, have drawn inspiration from STS and themselves become its drivers.

The practice of care

Just as competence and care have been singled out as defining an inherent tension in medicine, this distinction may apply also to medical anthropologists; one could characterise those with a penchant to STS as engaging with medicine's claim to competence, while others, often equally dedicated to a social critique of science and society, have undertaken research into what is here termed 'the practice of care'. Care is here understood in a broader sense than in medicine, where it encompasses general practice, the caring professions and home-based medicine. Medical anthropologists have long gone beyond medical sociology's concern with the patient-practitioner dyad and have linked the sociology of situated knowledge production to situated caring practices as an aspect of sociality more generally

In Europe, early work into the practice of care centred on migration and health, a topic that trans-cultural psychiatry had earlier addressed in its narrower formulation of migration and mental health. The concept 'culture-bound syndrome' is a legacy from trans-cultural psychiatry that even today has not yet been entirely buried, as it resurfaces in undergraduate essays and colleagues' well-meaning comments. Medical anthropologists, however, have long discarded this concept from their analytic toolkit, not least as it medicalises unusual forms of conduct and thereby exoticises other peoples and 'ethnic' immigrants. Early criticisms (e.g. van Dijk 1998 [1989]) have more recently been reconceptualised and reformulated (e.g. Fassin 2000) and generalized into a critique of culturalism (e.g. Olivier de Sardan 1999). Yet, as Fainzang (2007 [2005]) notes, one need not do away with the concept of culture entirely, as this concept encompasses historical and political dynamics that cannot be ignored even in the study of micro-social settings such as the clinical encounter. The increasing mobility of labour forces into and within the European Union has kept the theme of migration as relevant as ever (e.g. Hüwelmeier and Krause 2011), also in studies on other continents (Napolitano 2003), while research into mental health has since seen many turns in theoretical orientation, of which one only will be discussed below.

The practice of care has also been a longstanding concern for medical anthropologists working in the context of development studies, which in Europe often involves Africa. It has been researched within the anthropology of pharmaceuticals, hospital ethnographies, patient agency, and the like, sometimes in close collaboration with South-Saharan colleagues (e.g. Van der Geest and Whyte 1988; Whyte, van der Geest, and Hardon 2003). The formulation of social pragmatism has gained distinctiveness, particularly in Copenhagen (e.g. Whyte 1997), but not exclusively (e.g. Benoist 1996; Geissler and Prince 2010). Research within this social pragmatic framework warrants long-term fieldwork, such as yearly returns to one's second home in Africa over decades. It is this sort of research that can account for the increasing problem of chronic conditions in ethnographies that holistically present the villagers' daily life (see Whyte, this issue).

The practice of care has also been of primary importance for research on AIDS, which, as Mattes (this issue) demonstrates, has steadily been integrating new research agendas. Where research into sexual networks and the prevention of HIV infection marked the early days, research now centres on anti-retroviral treatment, the social inequalities that access to it generates and upholds, and moral issues ensuing from the chronic nature of the condition. Although medical anthropologists became involved in this theme later than epidemiologists, whose concept of risk was instantly contested (e.g. Frankenberg 1993), concerns relating to the care of AIDS in a world of increasing uncertainty and risk have left a lasting mark in the field (e.g. Dilger and Luig 2010).

Finally, the practice of care has pre-occupied researchers who responded creatively to the era of postmodernism by initiating ethnographic research at home in urban and modern settings (e.g. Fainzang 2000 [1989]; van Dongen 2004), at a time when elsewhere it engendered excessive navel gazing. Medical anthropology at home, MAAH, constitutes perhaps one of the most distinctively European endeavours. This is said in awareness that the association's members pursue diverse research interests. Historically, medical anthropology evolved in close association with folklore studies in Italy and Spain (e.g. Sepilli, this issue; Comelles 1996), much in contrast to France and German-speaking countries, and arguably Lithuania, Latvia and Finland (Vaskilampi 1994), where social/cultural anthropologists and folklorists have tended to maintain a guarded distance. Some MAAH members have since adopted a more political economic perspective, others a distinct pragmatic stance, and yet others combine an STS analysis within controversies over citizenship, in the light of the increasingly legalistic developments of health care.

From the body politic to the body

The 'body politic', rather than the body, has been central to medical anthropology since the early days. Witchcraft, spirit possession and childbirth, themes that typically concerned women, tended to be analysed in view of gendered power relations within the body politic, although the authors themselves did not use the term. To be sure, witchcraft and spirit possession were used to frame social processes in ways that medical anthropologists contested, by highlighting either that people's health concerns are generally not as exotic as supposed (see previous section on social pragmatism) or that the plight experienced by every bewitched is actually a quite common human condition (see next section on phenomenology). Witchcraft suspicions are often directed at women, as in the currently rampant killings among the Sukuma (Mesaki 2009) – often women to whom men are vitally indebted, e.g. the paternal aunts, whose departure into another lineage provides the men the cattle they depend on for founding a homestead (Stroeken 2010: 145).

Likewise, it is through a medical anthropological lens, combined with a grounding in gender studies, by which I.M. Lewis' functionalist explanation of so-called peripheral 'spirit possession' as a socially disregarded, hidden, mostly female practice, has been revised. Well-known is Boddy's (1989) compelling ethnography on women's life worlds and the intricate interdependencies of the segregated gendered spaces in the Sudan, but it often goes unacknowledged that De Martino (2005 [1961]) had earlier brought phenomenology into the analysis of spirit possession in the tarantula dance of southern Italy. De Martino focused on how the religious experiences of the politically subaltern interrelated with their historically grown socio-economic condition (see Pandolfi and Bibeau 2007 [2005]: 127–8), while the gender dimension has since been foregrounded, particularly in studies that bring the body political into focus (e.g. Behrend and Luig 1999; van Dijk, Ries, and Spierenburg 2000).

While spirit possession generally has been discussed in the anthropology of religion (e.g. Sax 2002; Gellner 1994), it has increasingly become a topic of medical anthropological research, as spirit possession practices have been peculiarly resistant to the encroachment of medical professionals in pluralist health fields. In this volume, for example, Bindi accounts for the body political through a focus on the bodily experience of becoming possessed. Bindi thereby expands on Pandolfi's early work (e.g. 1991), which started with narrative analyses but led on to an account of embodied memory and emotion, and is exemplary in illustrating how the medical anthropological concern with the 'body political' has increasingly shifted to the 'embodied political', and thereby to an increased focus on the body.

Finally, the early research on midwifery and birthing can also be appreciated as research into the gendered body politic, although initially it was not expressed in those words. Early literature on childbirth often combined a xenophilia for other peoples with a critical view on medicalised practices in North America and Europe. It was often biomedical professionals who took other people's practices as a source of inspiration to effect a change in modern medical birthing facilities and to instil a renewed appreciation of the homebirth. It appears that precisely this heterotopic literature on childbirth has put women in industrialised countries under additional stress and, particularly in cases of complications, aggravated self-doubt (Kneuper 2003). More recent literatures emphasize that providing adequate maternity services remains a body political issue, not merely in the Southern hemisphere (e.g. Unnithan 2004), but also in Eastern Europe (Putnina 1999).

Witchcraft, spirit possession and childbirth thus all bring into focus the cross-culturally observed precarious position of women. Research into the body politic has since become more diverse, whereby globalization, medico-scapes, the pharma-industrial complex, substance abuse, and violence are discussed alongside medicine as affected by nationalism and the moral economies of ethnicity (e.g. Samuelsen and Steffen 2004; Lambert and McDonald 2009; Ecks and Harper in press).

Where medical anthropology in its early days emphasized that medical problems were not primarily to be found in the body but in the social, one can observe a re-appraisal of the empiricist's biological body in *Beyond the body proper* (Lock and Farquhar 2007): a social critique need not imply disregard for any bio-scientific findings. The dissected, dead and static body that social scientists know as the biomedical one is only one of the many described in the bio-sciences, where research on living bodies ranges from the study of electro-communicative fibres of the connective tissue to the ecologies of diverse genomic micro-organismic populations of the gut. However, some medical anthropologists have gone yet another step further by highlighting that the biological body as a fine-tuned project-in-the-making implicates the body political.

Where Bindi (this issue) highlights how the body politic of competing medical treatment choices is experienced in the individual bodies of patients, Pizza (this issue) studies how the (non-)collaboration of diverse medical specialists, and their body political pre-occupation with trying to save time, leads to an increased likelihood of diagnoses of one of the most feared bodily conditions, dementia due to Alzheimer's. Gramsci's notion of 'second nature', which Pizza explains in a very accessible prose, allows for a micro-social, or what Gramsci calls a 'molecular' analysis, which simultaneously accounts for the bodily and body political. In his analysis of how early diagnoses of Alzheimer's are produced, Pizza makes the Gramscian concept of 'second nature' appear less disembodied than Bourdieu's concept of habitus and also less deterministic than a Foucauldian inscription of the state on the individual. 'Second nature's' constant reworking forfeits social determinism: in endlessly many minute micro-social situations, it allows the actors to have self-reflection, and to rework it accordingly.

The psy-dimension, the senses, emotions and aesthetics

The *Anthropology of the body* was edited by an ethnomusicologist (Blacking 1977a) and, therefore, has not generally been on introductory reading lists of medical anthropology.[1] It discusses human activities that are bodily in yet another way: from a social anthropological perspective that does not disregard biological anthropological considerations entirely.[2] It asks, for instance, how making chamber music can generate moments that are emotionally extremely rewarding, as for instance moments when the individual experiences a dissolution in sound and a melding with the group as a whole (Blacking 1977b). Blacking's perspective takes as its starting point group-coordinated human activity that is aesthetically pleasing and links it to bodily, sensorial and emotionally-felt processes of sociality. His approach meets well with the current medical anthropological interest in the bodily. In contrast to the empiricist scientist's gaze onto an individual's bounded biological body-enveloped-by-skin, the focus is on people's aesthetically pleasing experiences of bodily interaction within a group.

Medical anthropologists have variously drawn on insights from symbolic anthropology, psychological anthropology and ethno-psychoanalysis, but it would appear that the most important source of intellectual stimulation on the illness-generating social processes, within which the emotional is inflected, comes from yet another direction both in early and current medical anthropological research. This is phenomenology. Today, medical anthropologists on both sides of the Atlantic draw on Merleau-Ponty, whom Thomas Csordas (e.g. Csordas 2002) put centre stage within the field, importantly, by including Bourdieu's concept of the *habitus* into the discussion of the pre-objective. However, many thinkers had propelled this philosophical movement already in the first part of the last century (among them the above-mentioned Ernesto de Martino). Phenomenology also inspired many physicians who initiated foundational research for medical anthropology. Among them are Franco Basaglia (1924–1981), whose achievements within the Italian anti-psychiatry movement transcended national boundaries (Pandolfi and Bibeau 2007 [2005]: 124–7), as well as F.J.J. Buytendijk in the Netherlands and Joachim Sterly in German-speaking countries (mentioned in the introductory essays), and it continues to guide medical doctors' anthropological explorations (e.g. Martínez-Hernáez 2000).

The contributions to this volume by René Devisch and his former student Koen Stroeken both take a phenomenological approach: Devisch to divinatory practices among the Yaka in Zaire, Stroeken to witchcraft accusations among the Sukuma in Tanzania. Yet each engages with phenomenology in a different way: Devisch emphasizes the matrixial and attends to symbols relevant to psycho-analytical reflection, while Stroeken's phenomenology insists on the validity of empirical observation, particularly among rural 'peasant intellectuals.'

Stroeken's paper also contains an interesting side note, which is that the emotional is often manipulated by the botanical. It is interesting that in the context of exploring the psy-dimension and aesthetics that ethnobotanical research resurfaces as relevant, as it has the potential to assist medical anthropologists in better understanding therapeutic techniques of emotional manipulation and modulation. As Stroeken (2010: 26) notes elsewhere, the plant mixtures that healers administer to their clientele always involve one plant acting as a connector, which may trigger a sensory shift that brings with it a transformation of the patient's emotional states. The plant that is credited with effectiveness typically is a fresh and often aromatic plant (on the connection between scent and sacred, see Lefevre, Randriahasipara, and Velonandro 2008: 103). Scent has a certain physicality (Parkin 2008), and not only the meanings it evokes but its very physicality is likely to affect the physicality of the emotional.

There is an unexamined, yet nevertheless widely accepted, assumption that empirical knowledge derives from indiscriminate experimentation in terms of trial and error. Recent phenomenologically-oriented research points to more intelligent bodies that are primarily practical, that, when they project themselves with intentionality into the world, have the ability and propensity to learn from and test the world in non-random ways. These approaches integrate the psy-dimension into the internal bodily-felt as well as into observable human interaction. They thereby aim to overcome the constructed division of the subjective from the objective. Just as it is important to locate the sensorial not in the person but in the interpersonal interrelation (Chau 2008), the emotional is best conceived as a process that arises in the interrelational.

A focus on the physicality of the emotional in human interaction, which Devisch does in all his writings, rather than speculating about the individual's interiority, calls for a paradigm shift. Devisch reminds us that the psy-dimension can only be comprehended by doing away with the artificiality of studying either semantics (meaning/knowledge) or pragmatics (doing/action).

Accordingly, the Word and the World are not as separate as social and natural scientists currently construct them to be. Latour (2000) speaks of a 'realistic realist' stance (with interesting implications for medical anthropology, see Hsu 2010). A realistic realist's ethnographic fieldwork of individually-felt, emotionally-rewarding, coordinated human interaction, with a focus on its processual aspects, is likely to transcend the limitations of actor-network theory that arise from locating agency in the actor instead of the interaction. Perhaps, such a medical anthropologist's realistic realist's perspective has the potential to make important inroads towards a more general anthropological understanding of the psy-dimension as it resonates with social and other environmental workings.

By way of conclusion

The practice of care – at home and in developing contexts, in applied and less applied settings – is certainly one of the core themes researched in Europe. The focus on practice (and knowledge as an aspect of practice) is probably best explained in light of the centrality that practice theory has enjoyed in social anthropology. The focus on care – in its broadest sense – reflects, no doubt, a critical stance, which many medical anthropologists share, towards impersonal, confidence-diminishing, disempowering, if not illness-inducing medical care in bureaucratic institutions.

The sections on the body politic and the psy-dimension highlighted the field's debt to anti-hegemonic social movements, such as the feminist and the anti-psychiatry movements. As the section about the body politic shows, the social practices that in anthropology are discussed under the rubrics of witchcraft, spirit possession and childbirth have much enhanced our understanding of the precarious position of women in many societies. In a similar vein, the phenomenological documentation of how bodies are experienced in aesthetically coordinated interaction promises to provide a fruitful mode of appreciating the psy-dimension and render obsolete excessive reliance on often surprisingly ethnocentric speculations about an individual's 'psychology'. Medical anthropological research of this kind, which acknowledges the debt to the Cartesianism built into our thinking but simultaneously engages in field-based inquiries that relativise this dichotomy, is likely to provide a framing of general relevance for future research that may transcend its disciplinary foundations.

Acknowledgements

The author thanks all the authors and vignette writers to this issue who took the time to comment on her earlier drafts. She also thanks Josep Comelles, Alice Desclaux, Suzette Heald, Helle Johannessen, Murray Last, Anne-Lise Middleton, David Parkin and Martine Verwey for clarifying specific points. Finally, special thanks go to Caroline Potter.

Conflict of interest: none.

Notes

1. This is so although Murray Last, Gilbert Lewis and Vieda Skultans have a paper in this ASA volume, as they do in Loudon (1976) and Littlewood (2007).
2. For a more recent, similarly creative attempt to overcome the widespread 'aversion' among medical anthropologists against biological anthropology, see Parkin and Ulijaszek (2007).

References

Barth, Fredrik. 1975. *Ritual and knowledge among the Baktaman of New Guinea*. Oslo: Universitets-forlaget.

Behrend, Heike, and Ute Luig, eds. 1999. *Spirit possession, modernity and power in Africa*. Oxford: James Currey.

Benoist, Jean, ed. 1996. *Soigner au pluriel, Essais sur le pluralisme medical*. Paris: Karthala.

Berg, Marc, and Annemarie Mol, eds. 1998. *Differences in medicine: Unraveling practices, techniques and bodies*. Durham: Duke University Press.

Blacking, John, ed. 1977a. *The anthropology of the body*. London: Academic Press.

Blacking, John, ed. 1977b. Towards an anthropology of the body. In *The anthropology of the body*, 1–28. London: Academic Press.

Boddy, Janice. 1989. *Wombs and alien spirits: Women, men, and the Zar cult in northern Sudan*. Madison: University of Wisconsin Press.

Chau, Adam. 2008. The sensorial production of the social. *Ethnos* 73, no. 4: 485–504.

Comelles, J.M. 1996. Da superstizioni a medicina popolare: La transizione da un concetto religioso a un concetto medico. *AM. Rivista Italiana di Antropologia Medica* 1–2: 57–8.

Csordas, Thomas. 2002. *Body/meaning/healing*. Basingstoke: Palgrave Macmillan.

De Martino, Ernesto. 2005 [1961]. *The land of remorse: A study of southern Italian tarantism*. Trans. D.L. Zinn. London: Free Association Books.

Dilger, Hansjörg, and Ute Luig, eds. 2010. *Morality, hope and grief: Anthropologies of AIDS in Africa*. Oxford: Berghahn.

Ecks, Stefan, and Ian Harper. In press. 'There is no regulation, actually': The private market for anti-TB drugs in India. In *When people come first: Anthropology, actuality and theory in global health*, eds. J. Biehl and A. Petreyna. Durham, NC: Duke University Press.

Fabrega, H. 1975. The need for an ethnomedical science. *Science* 189, no. 4207: 969–75.

Fainzang Sylvie. 2000 [1989]. *Of malady and misery: An Africanist perspective of illness in Europe*. Amsterdam: Het Spinhuis.

Fainzang, Sylvie. 2007 [2005]. Medical anthropology in France: A healthy discipline. In *Medical anthropology: Regional perspectives and shared concerns*, eds. F. Saillant and S. Genest, 89–102. Malden: Blackwell.

Fassin, D. 2000. Les politiques de l'ethnopsychiatrie. La psyché africaine, des colonies africaines aux banlieues parisiennes. *L'Homme* 153: 231–50.

Frake, C.O. 1961. The diagnosis of disease among the Subanun of Mindanao. *American Anthropologist* 63, no. 1: 113–2.

Frankenberg, Ronnie. 1993. Risk: Anthropological and epidemiological narratives of prevention. In *Knowledge, power, and practice: The anthropology of medicine and everyday life*, eds. Shirley Lindenbaum and Margaret Lock, 219–42. Berkeley: University of California Press.

Franklin, Sarah. 1997. *Embodied progress: A cultural account of assisted conception*. London: Routledge.

Geissler, Paul Wenzel, and Ruth Jane Prince, eds. 2010. *The land is dying: Contingency, creativity and conflict in western Kenya*. Oxford: Berghahn.

Gellner, D.N. 1994. Priests, healers, mediums and witches: The context of possession in the Kathmandu Valley, Nepal. *Man* 29, no. 10: 27–48.

Gellner, Ernest. 1974. *Legitimation of belief*. Cambridge: Cambridge University Press.

Good, Byron J. 1977. The heart of what's the matter: The semantics of illness in Iran. *Culture, Medicine and Psychiatry* 1, no. 1: 25–58.

Horton, Robin, and Ruth Finnegan, eds. 1974. *Modes of thought: Essays on thinking in Western and non-Western societies*. London: Faber.

Hsu, Elisabeth. 2010. Introduction: Plants in medical practice and common sense. In *Plants, health and healing: On the interface of ethnobotany and medical anthropology*, eds. Elisabeth Hsu and Stephen Harris, 1–48. Oxford: Berghahn.

Hüwelmeier, Gertrude, and Kristine Krause, eds. 2011. *Traveling spirits: Migrants, markets and mobilities*. London: Routledge.

Kneuper, Elsbeth. 2003. Die 'natürliche Geburt' – eine globale Errungenschaft? In *Medizin und Globalisierung: Universelle Ansprüche – lokale Antworten*, eds. Angelika Wolf and Viola Hörbst, 107–28. Münster: Lit.

Lambert, Helen, and Maryon McDonald, eds. 2009. *Social bodies*. Oxford: Berghahn.

Latour, B. 2000. When things strike back. A possible contribution of science studies to the social sciences. *British Journal of Sociology* 51: 107–23.

Latour, Bruno, and Steve Woolgar. 1979. *Laboratory life: The construction of scientific facts*. Beverly Hills: Sage.

Lefevre, G., M.L. Randriahasipara, and Velonandro. 2008. Les fleurs de Tselatra. Symbolique des flerus et condition humaine chez un poète malgache du début du XXe siècle. *Etudes Océan Indien* 40–41:101–37.

Littlewood, Roland, ed. 2007. *On knowing and not knowing in the anthropology of medicine*. Walnut Creek: Left Coast Press.

Lock, Margaret, and Judith Farquhar, eds. 2007. *Beyond the body proper: Reading the anthropology of material life*. Durham: Duke University Press.

Loudon, J.B., ed. 1976. *Social anthropology and medicine*. London: Academic Press.

Martínez-Hernáez, Angel. 2000. *What's behind the symptom? On psychiatric observation and anthropological understanding*. Amsterdam: Harwood.

Mesaki, Simeon. 2009. The tragedy of ageing: Witch killings and poor governance among the Sukuma. In *Dealing with uncertainty in contemporary African lives*, eds. Liv Haram and C. Bawa Yamba, 72–90. Uppsala: Nordic African Institute.

Mol, Annemarie. 2002. *The body multiple: Ontology in medical practice*. Durham: Duke University Press.

Napolitano, Valentina. 2003. *Migration, mujercitas, and medicine men: Living in urban Mexico*. Berkeley: University of California Press.

Needham, Rodney. 1975. Polytheic classification: Convergence and consequences. *Man* 10, no. 3: 349–69.

Olivier de Sardan, Jean-Pierre, ed. 1999. Editorial *Bulletin de l'APAD* 17: 2–3.

Pandolfi, Mariella. 1991. Memory within the body: Women's narratives and identity in a southern Italian village. In *Anthropologies of medicine*, eds. Beatrice Pfleiderer and Gilles Bibeau, 59–65. Heidelberg: Vieweg.

Pandolfi, Mariella, and Gilles Bibeau. 2007 [2005]. Suffering, politics, nation: A cartography of Italian medical anthropology. In *Medical anthropology: Regional perspectives and shared concerns*, eds. F. Saillant and S. Genest, 122–41. Malden: Blackwell.

Parkin, David. 2008. Wafting on the wind: Smell and the cycle of spirit and matter. In *Wind, life, health: Anthropological and historical approaches*, eds. Elisabeth Hsu and Chris Low, 37–50. Oxford: Blackwell.

Parkin, David, and Stanley Ulijaszek, eds. 2007. *Holistic anthropology: Emergence and convergence*. Oxford: Berghahn.

Polanyi, Michael. 1998 [1958]. *Personal knowledge: Towards a post-critical philosophy*. London: Routledge.

Putnina, Aivita. 1999. Maternity services and agency in post-Soviet Latvia. PhD thesis in Social Anthropology, University of Cambridge.

Samuelsen, Helle, and Vibeke Steffen, eds. 2004. *The relevance of Foucault and Bourdieu for medical anthropology: Exploring new sites*. Special Issue, *Anthropology & Medicine* 11, no. 1: 3–105.

Sax, William. 2002. *Dancing the self: Personhood and performance in the Pandav Lila of Garhwal*. New York: Oxford University Press.

Stroeken, Koen. 2010. *Moral power: The magic of witchcraft*. Oxford: Berghahn.

Unnithan-Kumar, Maya. 2004. *Reproductive agency, medicine and the state: Cultural transformations in childbearing*. Oxford: Berghahn.

Van der Geest, Sjaak, and Susan Reynolds Whyte, eds. 1988. *The context of medicines in developing countries: Studies in pharmaceutical anthropology*. Dordrecht: Kluwer.

Van Dijk, Rob. 1998 [1989]. Culture as excuse: The failures of health care to migrants in the Netherlands. In *The art of medical anthropology: Readings*, eds. Sjaak van der Geest and Adri Rienks, 243–50. Amsterdam: Het Spinhuis.

Van Dijk, Rijk, Ria Reis, and Maja Spierenburg, eds. 2000. *The quest for fruition through ngoma: The political aspects of healing in southern Africa*. Oxford: James Currey.

Van Dongen, Els. 2004. *Worlds of psychotic people: Wanderers, 'bricoleurs' and strategists.* London: Routledge.

Vaskilampi T. 1994. Alternative medicine in Finland – an overview on the role of alternative medicine and its research in Finland. In 1994. *Studies in alternative therapy,* eds. Johannessen, Helle, Soren Gosvig Olesen, Jorgen Ostergard Andersen, and Erling Høg, Vol. 1. 40–5. Odense: Odense University Press.

Whyte, Susan Reynolds. 1997. *Questioning misfortune: The pragmatics of uncertainty in eastern Uganda.* Cambridge: Cambridge University Press.

Whyte, Susan Reynolds, Sjaak van der Geest, and Anita Hardon. 2003. *Social lives of medicines.* Cambridge: Cambridge University Press.

Chronicity and control: framing 'noncommunicable diseases' in Africa

Susan Reynolds Whyte

Department of Anthropology, University of Copenhagen, Copenhagen, Denmark

This paper proposes a way of framing the study of 'noncommunicable diseases' within the more general area of chronic conditions. Focusing on Africa, it takes as points of departure the situation in Uganda, and the approach to health issues developed by a group of European and African colleagues over the years. It suggests a pragmatic analysis that places people's perceptions and practices within a field of possibilities shaped by policy, health care systems, and life conditions. In this field, the dimensions of chronicity and control are the distinctive analytical issues. They lead on to consideration of patterns of sociality related to chronic conditions and their treatment.

In the last decade, the World Health Organization has sharpened its focus on Noncommunicable Diseases (NCDs) as problems of the global South as well as the global North. It emphasizes that low-income countries, where infectious diseases have been considered the main health issues, are now struggling with a 'double burden' of both communicable and noncommunicable disease. How can anthropologists contribute to a better understanding of this situation? This paper proposes a way of framing the study of 'noncommunicable diseases' in Africa, taking as points of departure the situation in Uganda, and the approach to health issues developed by a group of European and African colleagues over the years. It suggests a pragmatic analysis that places people's perceptions and practices within a field of possibilities shaped by policy, health care systems, and life conditions. In this field, the dimensions of chronicity and control are the distinctive analytical issues, for which earlier work has prepared the way. They lead on to consideration of patterns of sociality related to chronic conditions and their treatment.

Enquiry will focus on the extent to which a health condition involves a long-term relationship with medical authorities and continuing use of medicine. This is to take a pragmatic, even teleological, view that the meaning of a condition is heavily shaped by what people (policy makers, health professionals, affected individuals and their families) think they can do about it. An excellent example is the changing significance of HIV after the roll-out of antiretroviral therapy (ART). In Africa, HIV is becoming a chronic, less- (if not non-) communicable disease for those on ART, and it illustrates some of the key problems in addressing NCDs. Other examples come from interventions that have changed ideas about a wide range of chronic conditions under the rubric of disability. It is partly from research on the responses to HIV and disability that ideas emerge for future work on NCDs as chronic conditions in Africa.

Anthropology has a rich legacy of research on chronic conditions, which has been central for the development of methods, concepts, and theories in medical anthropology. It was long-lasting illnesses that prompted the development of concepts of therapeutic quests and therapy managing groups (Janzen 1978). Their duration made them suitable for the narrative approach to the telling of lives and suffering, in which one thing leads to another, and narrators try to find meaning and ways forward (Good 1994; Mattingly 1998; Skultans 2003). Theorizing illness identities (Estroff 1993) and biosociality (Rabinow 1996; Petryna 2002) revolved around chronic, not acute, conditions.

Looking to the specific chronic conditions that WHO is highlighting, there are also important ethnographic contributions: Mol's books on care for diabetes (Mol 2008) and knowledge of atherosclerosis (Mol 2002), Becker and Kaufman's (1995) studies of stroke, and the work of Mary-Jo Good and colleagues on cancer (Good et al. 1990, 1994). However, it is striking that most ethnographic research on cancer, cardiovascular conditions, and diabetes has been conducted in North America, Australia, and Europe. True, anthropologists have examined cultural differences in relation to these conditions; but just as was the case with research on disability until the 1990s (Whyte and Ingstad 1995, 4), most 'cross-cultural studies' have been carried out in the multi-ethnic settings of high-income countries with fairly good health care systems. Anthropologists interested in diabetes have studied minority communities in the global North (e.g. Schoenberg et al. 2005; Chowdhury, Helma and Greenhal 2000; Poss and Jezewski 2002; Borovoy and Hine 2008; Mendenhall et al. 2010) with particular attention to type 2 diabetes among Native Americans, among whom high prevalence is reported (Szathmáry 1994; Garro 1996; Sunday and Eyles 2001; Ferreria and Lang 2006). Much of this work has examined causal explanations and situations of post-colonial 'stress'.

The prospects suggested here for a research frame in Uganda and other African countries differ in several respects. There is less concern with causal attributions, except as they reflect different social positions and the spread of biomedical and global health ideas. Where earlier studies sought to identify cultural specificities in perceptions, this framing emphasizes the way different ideas are used (or not) in practice. It is more interested in 'actually existing health care systems' and the possibilities they offer for people dealing with chronic conditions. While all health care systems are composites (public/private, biomedical/alternative, specialist/ popular), African systems have particular characteristics (Turshen 1999). The public biomedical sectors are under-funded, poorly functioning, quite donor-dependent and poorly oriented to the control of chronic conditions, a fundamental that is hardly problematized in most anthropological approaches to long-term illness. In this situation, how do people try to control their conditions? The proposal is to trace social relations as resources for care and as affected by living with chronic conditions. Relations to health care providers and fellow sufferers are one place to start. Although biosociality, therapeutic citizenship, and not least therapeutic clientship, are relevant here, so are other relations to family, neighbors and colleagues not based on biomedicine.

'Noncommunicable disease' and chronic conditions

In global health discourse, the term NCD came into use in opposition to 'infectious' or communicable diseases (CDs). The old notion that an epidemiological transition would shift major causes of morbidity from CDs to NCDs was replaced by the recognition of the 'double burden', especially in African countries where malaria, tuberculosis and HIV have not disappeared with the increase in diabetes, cardiovascular diseases, and cancer (Marshall 2004). Now there is growing recognition among biomedical scientists that NCDs and CDs are often in close association rather than opposition (Young et al. 2009). For example, many forms of cancer are caused by infections (WHO 2011, 25–26), and treatment of HIV is implicated in the development of diabetes (Brown et al. 2005). Thus, NCDs and CDs may be overlapping and synergetic, rather than representing two contrasting patterns of disease.

In WHO policy, the boundary between disability and NCDs used to be sharper, when disability was seen mainly as permanent impairment of sensory or motor capacity. The new International Classification of Functioning, Disability and Health (ICF) adopted by WHO in 2001 broadens the definition to include all kinds of health conditions that affect functioning. 'The ICF thus "mainstreams" the experience of disability and recognizes it as a universal human experience. By shifting the focus from cause to impact it places all health conditions on an equal footing allowing them to be compared using a common metric – the ruler of health and disability' (WHO 2001). Inasmuch as diabetes, epilepsy, or the frailty of old age diminish the ability to participate in the social life of a given setting, they can be seen as disabilities. Still, the older conception of disability persists to some extent in the notion that the response to disability is rehabilitation in order to regain function, while control of NCDs is necessary to prevent deterioration. While chronic illnesses may be disabling, many disabilities would not be considered chronic illnesses because intervention is not necessary to prevent worsening. Simply put, as Irving Zola's informants explained, you can die from chronic illness, but not from disability (Zola 1982, 53).

It is evident from the initiatives of the WHO on Noncommunicable Diseases that the primary interest is in those common diseases that can be prevented with due attention to risk factors. The prototype NCDs are cardiovascular conditions (heart disease, hypertension, and stroke), cancer, chronic respiratory conditions, and type 2 diabetes (Daar et al. 2007). Epilepsy, sickle cell disease, and mental illness are also NCDs strictly speaking, but they are not so easily preventable. The NCDs that are targeted are those known as 'lifestyle diseases', associated with patterns of consumption and physical exercise that can be changed. The responsibility for intervening in these patterns, and also for medicating risk once it appears, is variously seen to lie with the individual, the family, the community, medical authorities, commercial interests, non-governmental organizations, and the state. The concept of 'lifestyle disease' has been criticized because it implies that people can choose how to live and stay healthy through proper self-discipline (Lupton 1995; Davison and Smith 1995). By contrast, seeing NCDs as related to changing 'life conditions', rather than styles, implies that larger forces of political economy are at play.[1]

Anthropologists do not have to accept the categories of WHO, which might be thought 'too biomedical'. But they are a part of the field of study, so medical anthropologists must consider them in relation to the perceptions of other actors. Preliminary explorations of chronic conditions in Uganda reveal that some health professionals are adopting the global health rhetoric of priorities and disease typologies. A Ugandan District Health Officer encouraged interest in cardiovascular disease and diabetes: 'These are the real Neglected Tropical Diseases', he asserted, invoking another category in global health research and intervention (Whyte, forthcoming a).

Rural, less educated people speak of the increase in 'the new sicknesses'. They mention *plesa* (high blood pressure), *sukari* (sugar or diabetes), *kansa* (cancer), *alusa* (ulcer), and *asama* (asthma). Alone, the terms for these new sicknesses suggest the spread of biomedical disease categories into popular domains; they are all vernacular variations of English words and concomitant meanings. Cosmopolitan people, especially members of the 'working class' (employees who earn regular salaries), are more likely to have received treatment for these sicknesses, and they are also more exposed to media and commercial products orientated to prevention and control.

Analytically, the notion of chronic conditions has been heuristically useful for anthropologists, because temporal persistence raises important conceptual possibilities.[2] It can encompass disability and HIV, mental problems, and the 'new sicknesses' ('noncommunicable' or 'lifestyle' diseases). Seeing them as part of the same field invites consideration of larger social and cultural processes. The point is not so much to distinguish biologically a viral disease like HIV from a 'lifestyle disease' like diabetes. Rather, it is to explore how they give meaning to one another by their similarities and contrasting possibilities of control, and how they are emerging and changing as social phenomena. 'Noncommunicable diseases' are communicable in a broader sense. Awareness of them is contagious; it spreads through experience with the health care system, through interactions with family and acquaintances, and through the media (Christakis and Fowler 2007; Montgomery et al. 2003). It is linked to the transaction of medicines and the bodily encounters with diagnostic equipment.

Control

Control or management can take many forms: prevention, therapy, palliation, risk limitation, rehabilitation, entitlement to support. The question is: how do available forms of control affect perceptions and practices? Previous work on disability has argued that the emergence of special education, the availability of assistive devices, rehabilitation programs, and movements for disability rights have created new meanings for sensory, motor, and intellectual impairments (Whyte and Ingstad 1995). In a country such as Uganda, the comprehensive category of disability (as opposed to blindness, deafness, slowness) has come into being through these forms of intervention. But chronic illnesses such as the prototype NCDs and HIV differ from the conventional disabilities in their closer relationship to biomedicine.

First of all, their diagnoses are more dependent on testing. Particularly in their early stages HIV, diabetes, cancer, and cardiovascular conditions are not definitively recognizable. Sometimes there are symptoms, and people may try various therapies, but screening is decisive for the biomedical treatment of these conditions. One of the most striking aspects of the Ugandan response to HIV has been the uptake of HIV testing and counseling. The message has gone home that only the test ('the machine' as people say) can reveal whether someone is HIV positive. Once ART became widely available together with other forms of social support (food rations, income generating possibilities), people were willing to test in large numbers. A positive test was an admission ticket for services that functioned, thanks to the influx of enormous resources for HIV.[3] Continued measurement of weight and CD4 counts became a part of many patients' awareness of their bodies and took on social significance (Meinert, Mogensen, and Twebaze 2009).

The same cannot be said of testing and monitoring for other chronic diseases. Screening is not widely carried out: at Ugandan public health facilities, blood pressure is seldom measured; equipment and supplies for checking glucose or cholesterol levels are lacking; the X-ray machine cannot be counted on. The current interest in technology within medical anthropology points to an obvious research issue around diagnostic equipment, its use, and consequences. A theme to explore is the relationship between availability and access to examination tools (weighing scales, blood pressure meters, and glucose tests), and the motivation to test on the part of both health workers and patients. As Hsu (this issue) writes, agency is thought to lie in the interaction between social actors and things.

This leads to a second point about control in lives with chronic illness: the presence of effective treatment programs is fundamental to the ways in which people manage their conditions. This may sound obvious, but it is sometimes forgotten, especially by anthropologists who are expected to discover 'cultural' factors in health perceptions and practices and in problems of 'adherence' to biomedical treatment. Paul Farmer (1999, 228–61) has argued emphatically that a user-friendly biomedical treatment system for long-term illness will be accepted, even if people think their illness was caused by witchcraft. Where daily medication is necessary over many years – and that is the case for many chronic conditions – one has to start by asking how easy it is to get the medicine.

'I Wish I Had AIDS,' was the title of a report on living with diabetes in Cambodia (Men 2007). Some Ugandans with diabetes express a similar sentiment. The response to AIDS demonstrated that it was possible, given sufficient resources and political will, to orient African health care systems to treatment of chronic diseases. But the same efforts have not been forthcoming for other long-term conditions. Whereas ARV drugs are being distributed for free in many African countries, people with diabetes have to buy their medicines and find money so they can take their drugs regularly (Aikins 2003; Awah, Unwin, and Phillimore 2008; Kolling, Winkley and Deden 2010).

Thirdly, control of chronic conditions involves measures other than taking pharmaceuticals therapeutically. More than is the case for impairments of sensory, motor, and intellectual capacities, conditions such as HIV, diabetes, and cardiovascular disorders evoke public health efforts of prevention and health promotion. Recommendations about changes in everyday patterns of sexuality, eating, drinking, and physical activity are often similar for people with diagnosed conditions and for those who would avoid them. Aside from research on sexuality in connection with HIV, a few ethnographic studies in middle- and low-income countries are addressing 'lifestyle' patterns, especially eating, in the light of the increase in NCDs (Wilson 2011).

Health education about HIV has been omnipresent, sponsored by government and donor programs, and by NGOs. In the absence of any comparable public health campaigns about preventing or living with 'lifestyle diseases', media and commercial interests nevertheless interpellate select publics. In this situation, an important theme for anthropological research is the way that prevention and treatment of 'the new sicknesses' illuminates differences – between richer and poorer, urban and rural, more and less educated. Not only can people with more resources purchase pharmaceuticals, they can also choose detoxifiers from 'Tianshi Products and Chinese Healthcare Culture' and cooking oil in bottles marked 'good for your heart' (rather than paying for a tiny amount at a time dipped from a large container of unknown origin). As in Europe, health consciousness and 'lifestyle' are becoming a mark of enlightenment and sophistication, but only for a fraction of the Ugandan population (Whyte forthcoming a).

If, as this framing suggests, the social existence of chronic conditions is linked to possibilities of control, it will be possible for researchers to follow their uneven development individually and collectively. Several African countries, including Uganda, have established offices for NCDs within their health ministries. Are there regular clinic days for diabetes and hypertension, as there are for mental illness and HIV? How does the introduction of the HPV vaccine for cervical cancer create awareness of this condition and for whom? One need not assume that control measures in themselves create or determine awareness and practice, but they certainly play an important part.

Time

Chronic conditions are by definition ones for which the time dimension is prominent, in several partly-related ways. One set of issues, already touched on above, concerns the historical transformations that are implicated in the increase in NCDs. Public health experts point to demographic ageing, 'rapid, unplanned urbanization' and 'the globalization of unhealthy lifestyles' (WHO 2011). Anthropologists studying illness perceptions find that lay people too see 'the new sicknesses' in the mirror of historical change: loss of 'traditional foods', increasing pollution, the stress of modern life, and poverty. While changing life conditions must be examined ethnographically, it is not political economic transformation as the cause of chronic illness that is suggested in this framing. Rather the questions revolve around the ways different kinds of people come to awareness of historical change, and how they make the links between changing conditions and chronic illness.

Illness time is of a different order than historical time. The temporal experience of illness may well be influenced by the health care system – a matter that should be explored. Time and control are linked in intricate ways, as Hsu has suggested. Diagnosis is often assumed to precede treatment, which in turn affects prognosis. Yet, in practice, treatment prospects influence diagnosis; prognosis and hope affect the labeling of illness: 'treatment, diagnosis and prognosis are bound into a complex web of interdependencies and fleeting temporalities' (Hsu 2005, 158).

In chronic conditions, there is often no singular acute beginning. As Bury (1982, 170) wrote: 'One of the most important features of chronic illness is its insidious onset. Non-communicable diseases do not "break-out" they "creep up".' In this, HIV resembles NCDs more than the communicable diseases with which it is usually classified. The timing of onset is complicated by the fact that the presence of disease, or risk of disease, may be determined by biomedical tests even before there are definite symptoms of illness. In high-income countries, therapeutic control of risk factors such as high values for cholesterol may be initiated even in the absence of biomedically defined disease, putting people in an ambiguous situation (Sachs 1995).[4] In Africa, people sometimes test HIV positive before they have any symptoms, but it is far less common that they are found at risk of cardiovascular disease, diabetes, or cancer without feeling ill, simply because screening is so rare.

A further aspect of timing concerns life-time and onset – the point in a biography when long-term illness sets in. Studies of disability show that the social significance of a condition is partly shaped by point of onset: a congenital impairment has different implications than one acquired late in life (Whyte and Ingstad 1995, 17). Typically, children are most affected by infectious diseases; mortality from malaria and respiratory infections is highest among under-fives. HIV is largely a sickness of young and middle-aged adults, while the prototypical NCDs are diseases of ageing.

Globally, the highest prevalence of type 2 diabetes is in the age group 40–59, but in 20 years it will be among those aged 60–79 (International Diabetes Federation 2009). Increasing longevity is one of the factors behind the 'epidemic' of chronic NCDs, and in the literature on ageing in the global North, an important theme is the medicalization of the later life course and the measuring and management of pathology (Cole 1992; Katz 1996; Kaufman, Shim and Russ 2004). It is striking that most anthropological research on ageing in Africa focuses on generational relations (Alber, Geest, and Whyte 2008; Geissler, Alber, and Whyte 2004) and on supportive care by and for elders (Ingstad et al. 1992). In this literature, old age itself is treated as a chronic condition, rather than a time of life when specific diagnoses require treatment. In contrast, the author is struck by the preoccupation of her Ugandan colleagues with the specific chronic health problems of their ageing parents, whom they bring to town in order to access treatment for cardiovascular conditions, diabetes, and cancer that is not available in the rural areas where their parents live. Perhaps researchers should attend more closely to generational relations around management of particular chronic illnesses of older people.

Seeing chronic conditions in relation to life-time leads to a final aspect of temporality: the experiential and processual time of familiarization and management. To live with a chronic condition does not imply stasis. There may be what feels at the time like a biographical disruption, a dramatic turning point, although this itself is an empirical question. Thereafter, chronicity has a course just as acute illness does. People learn to adjust their lives to restrictions of their bodies and (sometimes) of their treatment regimens. The condition becomes, if not normal, at least familiar. Yet even then, the course is not necessarily flat, straight, and consistent as the term chronicity might suggest. Flare-ups, acute episodes, improvements and deteriorations characterize chronic illness. Disability, depending on its nature, may be more even in its course, but rehabilitation, assistive devices, or changes in life conditions may also affect it. Most important, living with a chronic condition has a social dimension; it has implications for the individual within the household and family, as a patient, and possibly in relation to others with similar conditions.

Sociality

Issues of identity, biosociality and citizenship have been central in anthropological (and sociological) research on chronic conditions in recent years. In the global North, social scientists point to new forms of sociality concomitant with the phenomenal growth of biomedicine and its technology. Patient support and advocacy groups have diagnosis and/or treatment as criteria for inclusion. The biosociality argument is that social relations, and even identities, are formed on the basis of technologically mediated biological characteristics (Rabinow 1996). The related notion of biological citizenship (Petryna 2002; Rose and Novas 2005) points to the way people's rights and claims in a polity are formulated in terms of biological characteristics.

While these ideas are more relevant in high-income societies, where biomedical technology is developing rapidly and is widely available and accessible, they have also proved useful in posing questions about the global South. Researchers point to the AIDS response in Africa, which includes new kinds of sociality, attempts to change subjectivity, and a vital relationship between HIV positive people and the polities that provide their treatment. They note that understaffing of biomedical facilities has motivated the incorporation of 'expert clients' in care activities, another kind of social relationship based on diagnosis and treatment (Kyakuwa 2009). 'Therapeutic citizenship' is the term used to highlight the novelty of this situation (Nguyen 2005, 2010; Richey 2006) and to focus on changes facilitated by ART technology. Patient organizations are emerging for other conditions too, such as diabetes, cardiovascular disease and cancer.[5]

Illuminating as they are, the concepts of biosociality and therapeutic citizenship are most relevant as heuristic devices rather than descriptors of assumed reality. Even in societies classified as high-income, people are differentially touched and moved by biomedical technology. Membership in patient organizations is selective. Preoccupation with the medical aspects of one's condition characterizes only some of those affected. And if this is so in wealthier parts of the world, what part do the 'biopower' tendencies play in the lives of people with much poorer access to biotechnology? This question must be posed together with another, about the social relations of access to technologies.

Whether and how people access social or biomedical technologies is dependent on other conditions of existence. What is evident in the case of ART sociality is that connections and clientship are critical (Whyte et al., forthcoming). People who had succeeded in living with HIV emphasized the importance of knowing someone who helped them to test and to get onto a treatment program. They underlined their personal relationships to the health care providers who supplied them with ARVs. And the sociality that sustained them through sickness and on chronic treatment was the everyday moral economy of kinship, partnership, and friendship. Without that, they could hardly survive as biological citizens and therapeutic clients.

Patterns of sociality involve gender and generation. It has long been recognized that women shoulder much of the burden in chronic care, in Africa as elsewhere (Desclaux, Msellati, and Walentowitz 2009). Likewise, adult children are usually closely involved in care of their elderly parents, a pattern that will be more pronounced as longevity increases. What is less remarked upon is the role of men in providing money for care. In settings where expenses for treatment must often be met out of pocket, and where men generally have better access to cash than women, they must be mobilized to support the purchase of medicines, payment for tests, transport to health facilities, and perhaps buying special foods.

The notion of 'embodied sociality' (Muyinda 2008) emphasizes that the social embeddedness of people living with chronic conditions is evident in the state and capacities of their bodies. But it will also be important to explore the ways in which biomedical management creates tensions between individuality and sociality. Medication and recommendations about eating, alcohol consumption, rest, and physical activity focus on the individual. Particularly in respect to food, research on living with HIV suggests the difficulties of special feeding for one person in a household. However, it also demonstrates the way that gifts of nutritious food items to a chronically ill individual become modes of expressing concern and the amity of kinship (Whyte forthcoming b).

Conclusion

The study of chronic conditions such as diabetes and hypertension has a marked public health orientation. Concerted public health and commercial efforts to address 'lifestyle diseases' have only recently started to come into play, and it is possible to follow the emergence of a growing consciousness of these 'new sicknesses'. It is not simply that 'lifestyle diseases', as socially recognized phenomena, are spreading. They are being selected, appropriated and achieved, by certain kinds of people under the specific conditions of African health care. A pragmatic approach asks how differently situated people live with, and try to control, chronic conditions in a context of health care more orientated to acute health problems. It shares with public health the focus on meliorating practice, including use of medical technology, and a basic concern with equity.

Acknowledgements

Thanks to the editors for suggestions and to the reviewer who wrote several pages of useful comments.

Conflict of interest: none.

The paper was presented at the conference 'Medical Anthropology in Europe' funded by the Wellcome Trust and Royal Anthropological Institute.

Notes

1. The WHO uses both concepts seemingly as complements to one another. It emphasizes structural determinants: 'negative effects of globalization, for example, unfair trade and irresponsible marketing, rapid and unplanned urbanization and increasingly sedentary lives' as well as poverty (WHO 2011, 33). And it stresses population-level structural interventions such as regulation. But it also recommends health education and raising public awareness about diet and physical activity, interventions that assume that knowledge will motivate people to change their lifestyles (WHO 2011, 56)
2. Chronic illness is itself a 'technical term' that came into biomedical discourse as a category in the inter-war years (Armstrong 1990).
3. PEPFAR (the US President's Emergency Plan for AIDS Relief) initially committed 15 billion dollars to fight AIDS over five years (2003–2008) and was renewed and expanded

in 2008. The Global Fund to Fight AIDS, Tuberculosis and Malaria, established in 2002, is another prominent funder of AIDS interventions.

4. 'Prescribing by Numbers' was the term Jeremy Greene (2006) adopted for the pattern of starting medication when test values reach a certain established (and adjustable) threshold.

5. The Uganda Diabetes and Hypertension Association, the Uganda Heart Research Foundation, and the Uganda Cancer Society are currently receiving support from Danida to form an NCD alliance in the country.

References

Aikins, A.D.G. 2003. Living with diabetes in rural and urban Ghana: A critical social psychological examination of illness action and scope for intervention. *Journal of Health Psychology* 8: 557–72.

Alber, E., S. van der Geest, and S.R. Whyte, eds. 2008. *Generations in Africa: Connections and conflicts*. Münster: Lit Verlag.

Armstrong, D. 1990. Use of the genealogical method in the exploration of chronic illness: A research note. *Social Science and Medicine* 30: 1225–7.

Awah, P.K., N. Unwin, and P. Phillimore. 2008. Cure or control: Complying with biomedical regime of diabetes in Cameroon. *BMC Health Services Research* [Online], 8. Available: http://www.biomedcentral.com/1472-6963/8/43

Becker, G., and S.R. Kaufman. 1995. Managing an uncertain illness trajectory in old age: Patients' and physicians' views of stroke. *Medical Anthropology Quarterly* 9: 165–87.

Borovoy, A., and J. Hine. 2008. Managing the unmanageable: Elderly Russian Jewish émigrés and the biomedical culture of diabetes care. *Medical Anthropology Quarterly* 22: 1–26.

Brown, T.T., S.R. Cole, X. Li, L.A. Kingsley, F.J. Palella, S.A. Riddler, B.R. Visscher, J.B. Margolick, and A.S. Dobs. 2005. Antiretroviral therapy and the prevalence and incidence of diabetes mellitus in the multicenter AIDS cohort study. *Archives of Internal Medicine* 165: 1179–84.

Bury, M. 1982. Chronic illness as biographical disruption. *Sociology of Health and Illness* 4: 167–82.

Christakis, N., and J.H. Fowler. 2007. The spread of obesity in a large social network over 32 years. *New England Journal of Medicine* 357: 370–9.

Chowdhury, A.M.M., C. Helman, and T. Greenhal. 2000. Food beliefs and practices among British Bangladeshis with diabetes: Implications for health education. *Anthropology and Medicine* 7: 209–26.

Cole, T.R. 1992. *The journey of life: A cultural history of aging in America*. Cambridge: Cambridge University Press.

Daar, A.S., et al. 2007. Grand challenges in chronic non-communicable diseases: The top 20 policy and research priorities for conditions such as diabetes, stroke and heart disease. *Nature* 450: 494–6.

Davison, C., and G.D. Smith. 1995. The baby and the bath water: Examining sociocultural and free-market critiques of health promotion. In *The sociology of health promotion: Critical analyses of consumption, lifestyle and risk*, ed. R. Bunton, S. Nettleton, and R. Burrows, London: Routledge.

Desclaux, A., P. Msellati, and S. Walentowitz. 2009. Women, mothers and HIV care in resource-poor settings. *Social Science and Medicine* 69: 803–6.

Estroff, S.E. 1993. Identity, disability, and schizophrenia: The problem of chronicity. In *Knowledge, power, and practice: The anthropology of medicine and everyday life*, ed. S. Lindenbaum and M. Lock, Berkeley: University of California Press.

Farmer, P. 1999. *Infections and inequalities: The modern plagues*. Berkeley: University of California Press.

Ferreira, M.L., and G.C. Lang, eds. 2006. *Indigenous peoples and diabetes: Community empowerment and wellness.* Durham NC: Carolina Academic Press.

Garro, L.C. 1996. Intracultural variation in causal accounts of diabetes: A comparison of three Canadian Anishinaabe (Ojibway) communities. *Culture, Medicine and Psychiatry* 20: 381–400.

Geissler, W., E. Alber, and S. Whyte, eds. 2004. Grandparents and grandchildren. *Africa* 74, no. 1: 1–20.

Good, B.J. 1994. *Medicine, rationality, and experience: An anthropological perspective.* Cambridge: Cambridge University Press.

Good, M.-J.D., B.J. Good, C. Schaffer, and S.E. Lind. 1990. American oncology and the discourse on hope. *Culture, Medicine, and Psychiatry* 14: 59–79.

Good, M.-J.D.V., T. Munakata, Y. Kobayashi, C. Mattingly, and B.J. Good. 1994. Oncology and narrative time. *Social Science and Medicine* 38: 855–62.

Greene, J.A. 2006. *Prescribing by numbers: Drugs and the definition of disease.* Baltimore: Johns Hopkins University Press.

Hsu, E. 2005. Time inscribed in space, and the process of diagnosis in African and Chinese medical practices. In *The qualities of time: Anthropological approaches*, ed. W. James and D. Mills, Oxford: Berg.

Ingstad, B., F.J. Bruun, E. Sandberg, and S. Tlou. 1992. Care for the elderly—care by the elderly: The role of elderly women in changing Tswana society. *Journal of Cross-Cultural Gerontology* 7: 379–98.

Janzen, John M. 1978. *The quest for therapy: Medical pluralism in Lower Zaire.* Berkeley: University of California Press.

Katz, S. 1996. *Disciplining old age: The formation of gerontological knowledge.* Charlottesville: University Press of Virginia.

Kaufman, S.R., J.K. Shim, and A.J. Russ. 2004. Revisiting the biomedicalization of aging: Clinical trends and ethical challenges. *The Gerontologist* 44: 731–8.

Kolling, M., K. Winkley, and M. von Deden. 2010. For someone who's rich, it's not a problem: Insights from Tanzania on diabetes health-seeking and medical pluralism among Dar es Salaam's urban poor. *Globalization and Health* 6: 1–9. http://www.globalizationandhealth.com/content/6/1/8.

Kyakuwa, M. 2009. More hands in complex ART Delivery? Experiences from the expert clients initiative in rural Uganda. *African Sociological Review* 13: 143–67.

Lupton, D. 1995. *The imperative of health: Public health and the regulated body.* London: Sage.

Marshall, S.J. 2004. Developing countries face double burden of disease. *Bulletin of the World Health Organization* 82: 556.

Mattingly, C. 1998. *Healing dramas and clinical plots: The narrative structure of experience.* Cambridge: Cambridge University Press.

Meinert, L., H.O. Mogensen, and J. Twebaze. 2009. Tests for life chances: CD4 miracles and obstacles in Uganda. *Anthopology and Medicine* 16: 195–209.

Men, C.R. 2007. 'I wish I had AIDS': Qualitative study on health care access among HIV/AIDS and diabetic patients in Cambodia. Research project supported by the European Commission (EuropeAID, Health/2002/045-809).

Mendenhall, E., R.A. Seligman, A. Fernandez, and E.A. Jacobs. 2010. Speaking through diabetes. *Medical Anthropology Quarterly* 24: 220–39.

Mol, A. 2002. *The body multiple: Ontology in medical practice.* Durham: Duke University Press.

Mol, A. 2008. *The logic of care: Health and the problem of patient choice.* Milton Park: Oxford: Routledge.

Montgomery, G.H., J. Erblich, T. DiLorenzo, and D.H. Bovbjerg. 2003. Family and friends with disease: Their impact on perceived risk. *Preventive Medicine* 37, no. 3: 242–9.

Muyinda, H. 2008. Limbs and lives: Disability, violent conflict and embodied sociality in Northern Uganda. PhD diss., University of Copenhagen.

Nguyen, V.-K. 2005. Antiretroviral globalism, biopolitics, and therapeutic citizenship. In *Global assemblages*, ed. A. Ong and S. Collier, London: Blackwell.

Nguyen, V.-K. 2010. *The republic of therapy: Triage and sovereignty in West Africa's time of AIDS*. Durham: Duke University Press.

Petryna, A. 2002. *Life exposed: Biological citizens after Chernobyl*. Princeton: Princeton University Press.

Poss, J., and M.A. Jezewski. 2002. The role and meaning of susto in Mexican Americans' explanatory model of Type 2 diabetes. *Medical Anthropology Quarterly* 16: 360–77.

Rabinow, P. 1996. Artificiality and enlightenment: from sociobiology to biosociality. In *Essays on the anthropology of reason*, ed. P. Rabinow, Princeton: Princeton University Press.

Richey, L.A. 2006. Gendering the therapeutic citizen: ARVs and reproductive health. CSSR Working Paper. Cape Town: University of Cape Town.

Rose, N., and C. Novas. 2005. Biological citizenship. In *Global assemblages: Technology, politics, and ethics as anthropological problems*, ed. A. Ong and S.J. Collier, Malden, MA: Blackwell.

Sachs, L. 1995. Is there a pathology of prevention? The implications of visualizing the invisible in screening programs. *Culture, Medicine and Psychiatry* 19: 503–25.

Schoenberg, N.E., E.M. Drew, E.P. Stoller, and C.S. Kart. 2005. Situating stress: Lessons from lay discourses on diabetes. *Medical Anthropology Quarterly* 19: 171–93.

Skultans, V. 2003. Culture and dialogue in medical psychiatric narratives. *Anthropology and Medicine* 10: 155–65.

Sunday, J., and J. Eyles. 2001. Managing and treating risk and uncertainty for health: A case study of diabetes among First Nations people in Ontario, Canada. *Social Science and Medicine* 52: 635–50.

Szathmáry, E.J.E. 1994. Non-insulin dependent diabetes mellitus among Aboriginal North Americans. *Annual Review of Anthropology* 23: 457–82.

Turshen, M. 1999. *Privatizing health services in Africa*. New Brunswick: Rutgers University Press.

WHO. 2001. *International classification of functioning, disability and health* (Geneva: WHO).

WHO. 2011. *Global status report on noncommunicable diseases 2010* (Geneva: WHO).

Whyte, S.R. forthcoming a. The publics of the 'New Public Health': Life conditions and 'lifestyle diseases' in Uganda. In *Changing states of public health in Africa: Ethnographic perspectives*, ed. R. Prince and R. Marsland. Athens, OH: Ohio University Press.

Whyte, S.R., ed. forthcoming b. *Second chances: Living with ART in Uganda*.

Whyte, S.R., and B. Ingstad. 1995. Disability and culture: An overview. In *Disability and culture*, ed. B. Ingstad and S.R. Whyte, Berkeley: University of California Press.

Whyte, S.R., M. Whyte, L. Meinert, and J. Twebaze. Forthcoming. Therapeutic clientship: Belonging in Uganda's mosaic of AIDS projects. In *When people come first: Anthropology and social innovation in global health*, ed. J. Biehl and A. Petryna. Durham: Duke University Press.

Wilson, C. 2011. 'Eating, eating is always there': Food, consumerism and cardiovascular disease. some evidence from Kerala, South India. *Anthropology and Medicine* 17: 261–75.

World Diabetes Federation. 2009. *Diabetes atlas*, 4th edn. Brussels: IDF.

Young, F., J.A. Critchley, L.K. Johnstone, and N.C. Unwin. 2009. A review of co-morbidity between infectious and chronic disease in Sub Saharan Africa: TB and diabetes mellitus, HIV and metabolic syndrome, and the impact of globalization. *Globalization and Health* 5. http://www.globalizationandhealth.com/content/5/1/9.

Zola, I. 1982. *Missing pieces: A chronicle of living with a disability*. Philadelphia: Temple University Press.

'I am also a human being!' Antiretroviral treatment in local moral worlds

Dominik Mattes

Institute of Social and Cultural Anthropology, Freie Universität Berlin, Germany

The experiences and practices of antiretroviral drug consumers in Tanzania are shaped by economic scarcity, limited state-provided social welfare, and fragile kinship-based solidarity. Embedding antiretroviral therapy (ART) in patients' 'local moral worlds' brings further existential dimensions to the fore that articulate closely with the priority the treatment acquires in their lives. An exemplary case study of a middle-aged HIV-positive man suggests that dignity, social recognition, and belonging may be of central interest and temporarily overshadow patients' concern for mere survival. A stronger focus on patients' moral concerns contributes to a better understanding of the complex dynamics that prevent HIV-positive people from becoming the 'pharmaceutical selves' that are promoted during treatment enrolment. Moreover, it is indispensable to account for the lived experiences of patients struggling with what too readily has been termed a 'chronic disease'.

AIDS in anthropology: from cultural constructions to moral concerns

While anthropology's engagement with HIV/AIDS began somewhat hesitantly (Heald 2003), the discipline today has delivered a staggering number of contributions on HIV/AIDS as a socio-cultural, political-economic and moral phenomenon, varying considerably in their subject of inquiry as well as theoretical positioning. Following Ingstad's (1990) early study on the incorporation of AIDS in Tswana healers' conceptions of disease causation, various authors have shown how AIDS was constructed as a consequence of witchcraft (Ashforth 2002) or as the violation of social norms (Wolf 2001), thus pointing to the significance of local processes of meaning-making in people's response to their suffering. Furthermore, such processes were shown to be closely tied to transnational political-economic dynamics that produce social inequalities and enhance the precariousness of life, and according to Farmer (1992) reflect systems of 'structural violence'. Fassin has since demonstrated compellingly that individual accounts of suffering are deeply imbued by national and global political history, thus highlighting the 'social determinations of individuals' biological fate' (Fassin 2007, xv).

Against a background of economic insecurity, increased human mobility (as people seek a better livelihood), and gradually eroding kinship networks, AIDS has been framed in narratives of moral decline and threatened social continuity. The reinvigoration of reformist (Muslim as well as Christian) tendencies has been explained as a result of these ruptures brought about by modernity and globalization (Beckmann 2009). Moralizing discourses of AIDS have long provided a plausible explanation for the destructive dimensions of the ravaging pandemic and fostered hope for a complete cure, which the biomedical discourse so inexorably precluded (Dilger 2007).

In an in-depth study on the disruptions caused by the increasing bereavements amongst the most productive age groups in Tanzania, Dilger (2005) documented the dilemmas of providing care for the sick, and of re-establishing their moral integrity (subsequent to their death) within kinship-based solidarity networks stressed to their breaking point. As is well-known (Whyte 1997), the solving of such dilemmas involves highly pragmatic stances that contest seemingly prescriptive frameworks of kinship relations and notions of belonging.

Since issues of stigma are closely interconnected to disclosure, it became crucial to relate people's silence about their health status to the practical options that their lifeworlds provided. A long-term study in Uganda pointed to the moral concerns of people whose practice of 'revealing and hiding one's HIV status [...] is inextricably tied to the wish to be recognized as a person who matters in everyday interactions even when one is dying' (Mogensen 2010, 62). It demonstrated how modes of talking about HIV/AIDS have gradually changed over the decades, until the availability of antiretroviral medicines (ARVs) 'started making people articulate their HIV status in the hope that this would prolong their lives' (Mogensen 2010, 74).

Indeed, talking openly became indispensable for gaining access to ARVs before their large-scale roll-out. And even with their increasing availability, the elicitation of testimonials remained a significant component of the construction of patients as self-responsible managers of their 'chronic' condition. The formation of 'therapeutic citizens' was considered to positively influence patients' adherence to antiretroviral therapy (ART) in West Africa (Nguyen et al. 2007), while the creation of 'responsibilized subjects' supported identity transformations of AIDS survivors in South Africa (Robins 2006). However, in Tanzania, articulations of adherence production within rapidly expanding mass-treatment programmes were found to build on rather disempowering educational procedures which contradicted the discourse on ART as a 'human right' that is 'claimed' by 'empowered citizens' (Mattes 2011).

Biomedical technology – and the ideological baggage that comes with it (cf. Hardon and Dilger 2011) – 'may not in fact transform identity and citizenship in all settings; other aspects of sociality, such as work, the family and community, may be far more important in people's management of their health and lives' (Gibbon and Whyte 2009, 97). Recent research in socio-politically and religiously diverse environments thus demonstrated the limited analytic applicability of 'new forms of citizenship' triggered through ART. Biomedically defined notions of responsibility proved insufficient in accounting for the struggles of HIV-positive persons, who 'try to act responsibly and make informed decisions, but [...] do this by taking into account not only their biological, but also their social, cultural, and economic context' (Beckmann 2010, 5). Rather than promoting ideas of therapeutic citizenship and rights-based sociality, ART has been shown to be incorporated into patients' 'ongoing attempts to manage, maintain, and negotiate existing socialites, such as clientship, kinship, and friendship' (Meinert, Morgensen, and Twebaze 2009, 208). The present paper intends to complement these insights by relating patients' practices and ideas regarding ART to their daily struggles of leading a moral life.

Experiences of ART in local moral worlds

Treatment adherence entails more than regular drug intake, balanced nutrition, and a healthy lifestyle. It also has to be understood as a socio-moral phenomenon. The author's examination of lived experiences in patients' social environments in urban Tanzania unveiled that a person's will and capability to take the medications as prescribed is deeply enmeshed with other existential human longings, which are at times attributed a higher priority than mere survival. The present contribution therefore focuses on morality, which the author equates to 'what is at stake in everyday experience in particular local worlds' (Whyte, Whyte, and Kyaddondo 2010, 97).

Individual emotions, interpersonal relations and collective experiences of poverty and political powerlessness constitute the 'local moral worlds' (Kleinman 1999, 87) of many HIV-sufferers in Tanzania. Permanent struggles to provide sufficient food for one's family, to cover school fees, the cost for relatives' burials, or the treatment of endlessly recurrent illnesses make up large parts of their reality. While some manage to maintain their treatment and to resume their economic and social activities despite such difficulties, in other cases, this chronic crisis provides an infertile ground for ART to thrive its miraculous blossoms of resurrection.

Below, a case study is presented that centres on the actions and experiences of a 40-year-old male ARV consumer, his family members, and a close friend. This patient's struggles to adhere to his biomedical treatment regimen illustrate a discrepancy between 'normative prescriptions' of ART programmes and individual 'everyday practice' (cf. Burawoy 1998, 5). Within the presented narrative, various dimensions come into play: collective experiences of economic scarcity, specific configurations of familial history, and intra-household tensions. Understanding this context of shifting social relations is crucial for understanding the patient's stance towards ART, and the priority he gave to a sense of dignity and social recognition as pivotal matters of human life.

The case study illustrates that labelling HIV/AIDS a 'chronic health condition' is problematic, as it implies a transition of a patient's life situations to 'continuity' in both physical and social terms. Framing HIV/AIDS in terms of chronicity accounts neither for people's ongoing struggles to negotiate the disruptions that the illness entails, nor for the difficult re-establishment of social relations and rebuilding of one's moral standing in an uncertain and unstable world (see Dilger 2010).

Methods and background of the study

This paper draws on 15 months of fieldwork (2008–2011) on ART in Tanga, a city on the Tanzanian Swahili coast where ART has been offered large-scale since late 2004. In total, the author conducted 130 interviews with health staff, patients, their relatives, practitioners of traditional medicine, and religious leaders, and engaged in participant observation in treatment centres, support group meetings, and patients' homes. Through the inclusion of multiple actors' perspectives on ART, the study aimed to trace how social interactions in medical institutions shaped the meanings that ARVs acquired in patients' everyday lives, the impact of ART on concepts of 'healing' beyond the biomedical paradigm, and its contribution to a changing public discourse on HIV/AIDS.

The selected case study resembles other patients' narratives in revealing never-ending struggles to make ends meet, disturbing episodes of experienced and apprehended discrimination, and the carefully constructed and strictly guarded secrecy around one's health status. Its particularity stems from the extraordinary salience of concepts such as morality, recognition, and belonging, which strongly shaped and were shaped by the protagonist's interrelations within his social surroundings. Thus, it exemplifies how closely intertwined experiences of illness and therapy can be with a patient's subjectivity, concept of self and existential moral longings.

'Your heart grows numb, it's better to die' – the shattering of familial solidarity

Joseph Mamboleo[1] began ART in 2005. At the time of the first interview in January 2009, he was unemployed (due to his health status he could not continue working as a truck driver) and unmarried (his 14-year-old daughter stayed with one of his sisters in another city). He lived with his sister Grace and his younger brother Isaac in their deceased mother's house, and like the other household members, he depended on the financial aid of his eldest brother Benjamin who lived about 300 km away.

While Benjamin's payments had not been sufficient for all household members to live on, Joseph's situation worsened further due to a family conflict about the deceased mother's house. Irene, another elder sister of Joseph's, intended to remove Grace, Isaac and Joseph from the house and to take over the property. A court case solved the dispute in Benjamin's favour, but since Benjamin understood Joseph to be on Irene's side, he excluded him from the payments that had constituted Joseph's sole income.

Joseph's situation became alarming after his sudden exclusion from the household's cash flow. He was barely able to provide himself with one meal per day, and while being aware of the possible consequences for his health, he had not taken his ARVs regularly for about a month prior to the interview. 'It happens that you swallow it and not even an hour and a half later you feel bile (Swahili: *nyongo*) coming up. You feel the urge to chew something in order to get rid of the bitterness in your mouth. [...] You wake up in the morning and you can't even walk because you're so dizzy!', he complained.

Joseph's description of the family conflict and his own miserable situation revealed feelings of isolation and deep disappointment: '[My brother's] cruel and bitter words have entered my heart. [...] Your heart grows numb, it's better to die than to wait for anyone to return your heart [courage] (*arudishe moyo wako*)', he stated. Grace had supposedly advised him to reconcile with Benjamin, but his emotional wound seemed too deep for this. Grace remembered Joseph's stubborn reply as, 'I don't want to bow before anyone, even if he has money!'

Joseph was given his own plate to eat from, and Grace would not wash any clothes for him. 'They tell us, using the same kitchen utensils is not infectious but we are afraid and that's [also] why I don't wash for him. His clothes are his, his alone', she explained. Joseph suspected that Grace had been jealous of his previously close ties with Benjamin, but he also related her hostile behaviour to her lack of education.

Isaac did not seem to engage in Joseph's affairs at all. 'He leaves the house in the morning at six and comes back at ten or eleven at night, he goes to his room and sleeps. He lives his very own life (*maisha yake ye' mwenyewe*)', Joseph explained. The social relations in his household were thus characterized by covert as well as open confrontations, and mere indifference. The destructive dynamics of the siblings' conflict blended with existent tendencies of ostracism, producing individualization amongst the family members and inflicting profound emotional pain on Joseph, for whom the disintegration of this solidarity network escalated to a visible threat to his very existence.

'Born from the same womb' – avoiding social intimacy

Joseph's experiences in his wider social environment were equally pervaded by a strong undertone of frustration and alienation. There was such an intense dynamic of stigmatization in his neighbourhood that another HIV-positive man had already strangled himself, he explained. Because Grace had spread the word, everybody knew about his HIV-infection too. He lamented that people would not 'invest' in him any more because they knew that 'this one is sick already'. Joseph vehemently refused to be transferred to his own neighbourhood's health centre when the management of ART became part of its portfolio. 'I told the doctor that at the taxi stand [in front of the hospital] there were many young men who went to school with me. [. . .] She forced me, but I told her "If you force me, I won't go and pick up the medicines there and neither will I come back here"', he remembered.

Blurring the distinction between friendship and business partnership, Joseph drew a direct link between social intimacy and the possibility of successfully pursuing an economic project in his neighbourhood, where he experienced intimacy as obstructive and 'strangeness' (*ugeni*) as empowering: 'Everyone is afraid of you and [. . .] thinks "you can't know where this person came from, [. . .] I better don't start a quarrel with him [. . .]." It's different with a person who knows you right from the cradle.'

To be emotionally hurt and morally disappointed were the facts that mostly affected Joseph. His concept of social intimacy which correlated with notions of dignity and belonging, led to his strong wish to leave his neighbourhood and live on his own. Several times he expressed his extreme disappointment about his closest kin's unexpectedly cruel behaviour. Deprived of solidarity and protection from a close biological relative, Joseph experienced an existentially consternating contradiction. He aimed for a clear-cut break with all close kin and searched for new supportive forms of sociality in the protective realm of strangeness.

'You never know where the risk is' – biosocial collectives and enclaves of concealment

In a far-away neighbourhood Joseph was 'adopted' into the business of renting out bicycles by his friend Michael, who was also undergoing ART. They shared the meagre daily income, cooked and ate together. They also reminded each other to take their medicines, cared for each other during sickness episodes, and paid strict attention that no one would learn about their HIV-infection. If the self-support groups that mushroomed all over the city are taken as examples of newly forming 'biosociality' (Rabinow 1996), Michael's and Joseph's 'household' could be considered a non-disclosing biosocial micro-collective. Although the other young men hanging out at the bicycle stand became Joseph's new peer group, he felt more closely connected to Michael due to their shared biomedical predicament.

While the two allies' microcollective at least provided for an income, it was still insufficient, and Joseph kept looking for further options. He recounted an odyssey through Tanga's economy of non-governmental social support. He asked the staff of an NGO about a job, but they shrugged their shoulders. He tried to join an HIV/AIDS support group which was providing access to a recently started micro-credit initiative, but he had neither the 10,000 Tanzanian Shillings (ca. US$6) to pay the admission fee nor the 1,000 TSh for the monthly contribution. Promises of health staff to help him find a job remained unfulfilled. They kept urging him to take his medicines instead. Joseph felt misunderstood and once more socially distanced: 'They always ask "Why did you stop taking the drugs?" And I tell them "I have stopped because I don't have enough food". Then they say "You must not stop! [. . .] Eat and take your medicine!" [. . .] Their thoughts and my thoughts as a user (*mtumiaji*) are different. [. . .] They don't believe me, [. . .] they don't understand me', he complained.

Joseph eventually found work as a night guard and was assigned night shifts at the property of an upscale local authority. When he confided to his ward that he was HIV-positive and what physical strain his job meant to him, he unexpectedly encountered sympathy. Joseph was occasionally allowed to enter the house during his shift in order to sleep. He was treated 'like his [the house owner's] child', Joseph said in expressing his feelings concerning the unfamiliar reaction to his disclosure.

With the help of NGO staff Joseph also finally managed to join the self-support group. He was waived the admission fee and monthly contribution. However, when the micro credit initiative started recruiting their beneficiaries, he faced the next hurdle. Like several others, he had no assets that could serve as a deposit in order to get a first credit of 100,000 TSh. It was private engagement again that enabled these destitute supplicants to form an own group and get credits: a well-known HIV-positive woman who functioned as a peer educator in various health facilities agreed to take on full responsibility for them. With his credit, Joseph bought two bicycles that he started renting out for his and Michael's benefit. The profit from this and Joseph's monthly salary made the situation look less sombre for a while. However, at the time of the third interview in June 2009, working day and night started to exact its toll on Joseph. He had fevers and skin sores, was very exhausted, and desperately longed for a week of rest.

Furthermore, the microcollective with Michael increasingly became a source of further uncertainty. Michael was involved in substance abuse, and Joseph blamed him for spending big parts of their money on drugs and women. He also recounted that Michael had not taken his ARVs for considerable periods of time and claimed that dozens of unopened pill boxes were hidden in his cupboard. However, he would not 'resign the contract' with his friend: 'He is a human and I am a human and we were all built without knowing where the [next] risk is lurking. You can't let this one down and say "This is your own risk" and then you find yourself falling just the same way.'

These episodes demonstrate the contingency pervading Joseph's life. He only managed to gain access to non-governmental support structures through the engagement of well-minded 'influential' persons and held on to the solidarity of the microcollective even after this had become a source of uncertainty itself. Besides the difficult task of finding an ally for the encounters with everyday life's chronicities, Joseph also deployed a defence mechanism of a more psychological type: he generated a concept of self based on notions of difference and moral integrity and teleologically reconceptualized his life.

'I won't die tomorrow' – defying the 'empty present'

Joseph would not ask Benjamin for forgiveness. He prioritized his moral standards over resuming his proper position within familial power-relations, which would have possibly re-opened the pathway to economic survival. His morally informed concept of self also emerged when he referred to a person whom his sister had invited to occupy his room after his departure. He did not intend to expel that person even though he was desperately longing for a place to rest: 'I'm educated, I understand [...]. I'm also a human being (*binadamu*), I'm not an animal', he reasoned. Constantly offsetting himself from his social surroundings, Joseph created a 'resistance identity' (Castells 1997, 8). It built on his level of education, which exceeded that of most people of his social environment. He also considered himself a deeply moral person 'by nature': 'I don't know how I was made. My soul hurts a lot. I see more satisfaction (*radhi*) in sleeping down here on the cement than to throw a person out of her bed. I was built in a very unique way (*kipeke yangu kabisa*).'

Another dimension of his resistance identity came to the fore when, questioned about his future prospects, he explicated a concept of all-encompassing justice based on a tragic experience: the death of his youngest brother. After Joseph had tested HIV-positive, this brother had shunned him. One day they had an argument, and the same day his brother left the house and was killed in an accident. 'With the help of the medicines', Joseph concluded, 'I will live exactly the amount of years that God has written for me. [...] You can count with my death tomorrow but *kumbe* (expression of astonishment): I won't die! [...] I still consider myself as the same as any other human being [...]'.

Joseph resolutely handled the subjunctivity of his own 'therapeutic emplotment' (Mattingly 1994, 811). Despite all social and economic uncertainty, his narrative projected a life demarcated by a place of his own to stay alive, with work, with enough food – and with ARVs. Instead of living in the 'empty present' (Davies 1997, 568f), he decidedly resisted yielding to the existential disruptions he had experienced and put all his effort into maintaining a sense of coherent temporality.

Conclusion: ART, subjectivity, and moral survival

Joseph seemed to place less value on his biological than on his moral existence. Although he was well aware that without ARVs he would not be able to achieve his goals, the drugs only figured as one theme among others. Dignity and social recognition constituted the core of his aspirations, and the question of belonging was more important to him than being a biomedically rational 'pharmaceutical self' (Dumit 2002, 126). Social intimacy and distance were constantly negotiated, and

Joseph seemed to always meticulously gauge how much indignity he could 'accept' in order to still keep up his self-esteem. Living on ART influenced this critical line of balance. The drugs aggravated situations of suffering from hunger, and their intake posed an additional danger of involuntary disclosure. But Joseph's central concern was not about living on ARVs without hunger, but about being accepted as a human being.

Biehl and Moran-Thomas (2009, 270) recommend 'to grasp the wider impact of how medical technologies are becoming interwoven in the very fabric of symptoms and notions of well-being' through a comparison of 'the ways such life forms are fundamentally altering interpersonal relations, domestic economies, and identity-making processes'. The presented case study suggests that one should not dismiss the reverse workings of existential personal and intersubjective experiences, which substantially shape individual moral and emotional worlds and provide the ground on which treatment adherence may – or may not – flourish. Individual moral concerns play an utterly important role in the life-long social and psychological management of living with HIV. ART patients' modes of acting, thinking, and feeling should be understood as 'rooted in a particular constellation that connects [...] political economy with collective experience and the individual's subjectivity' (Biehl, Good, and Kleinman 2007, 3). A reinforced effort to include this idea within the system of access to and control of antiretroviral mass treatment programmes, for instance through creating more space and time to assess individual patients' moral worlds, might enable struggling patients such as Joseph to better maintain their treatment over a lifetime.

Acknowledgements

The Fritz-Thyssen-Stiftung, the National Institute of Medical Research and Tanzania Commission for Science and Technology provided funding and ethical approval. The author thanks Hansjörg Dilger, Claire Beaudevin, the Arbeitskreis Medical Anthropology at the Freie Universität Berlin, and the editors and reviewers for their insightful suggestions. Finally, he expresses his great indebtedness to the people living with HIV in Tanga.

The paper was presented at the conference 'Medical Anthropology in Europe' funded by the Wellcome Trust and Royal Anthropological Institute.

Conflict of interest: none.

Note

1. All names in this paper are pseudonyms.

References

Ashforth, A. 2002. An epidemic of witchcraft? The implications of AIDS for the post-apartheid state. *African Studies* 61, no. 1: 121–43.

Beckmann, N. 2009. AIDS and the power of God: Narratives of decline and coping strategies in Zanzibar. In *AIDS and religious practice in Africa*, ed. F. Becker and W. Geissler, 119–54. Leiden: Brill.

Beckmann, N. 2010. Responsibilized citizens? Care of the self among HIV-positive people in Tanzania. Paper presented at the Irmgard Coninx Foundation's 13th Berlin

Roundtables on Transnationality. Social Science Research Center, Berlin, 1–5 December, 2010.

Biehl, J., B. Good, and A. Kleinman, eds. 2007. *Subjectivity: Ethnographic investigations.* Berkeley: University of California Press.

Biehl, J., and A. Moran-Thomas. 2009. Symptom: Subjectivities, social ills, technologies. *Annual Review of Anthropology* 38: 267–88.

Burawoy, M. 1998. The extended case method. *Sociological Theory* 16, no. 1: 4–33.

Castells, M. 1997. *The power of identity.* Oxford: Blackwell.

Davies, M. 1997. Shattered assumptions: Time and the experience of long-term HIV positivity. *Social Science & Medicine* 44, no. 5: 561–71.

Dilger, H. 2005. *Leben mit AIDS. Krankheit, Tod und soziale Beziehungen in Afrika.* Frankfurt a.M.: Campus.

Dilger, H. 2007. Healing the wounds of modernity: Salvation, community, and care in a Neo-Pentecostal church in Dar-es-Salaam, Tanzania. *Journal of Religion in Africa* 37: 59–83.

Dilger, H. 2010. Introduction. Morality, hope and grief: Towards an ethnographic perspective in HIV/AIDS research. In *Morality, hope and grief: Anthropologies and AIDS in Africa,* ed. H. Dilger and U. Luig, 1–18. New York: Berghahn.

Dumit, J. 2002. Drugs for life. *Molecular Interventions* 2, no. 3: 124–7.

Farmer, P. 1992. *AIDS and accusation: Haiti and the geography of blame.* Berkeley: University of California Press.

Fassin, D. 2007. *When bodies remember: Experiences and politics of AIDS in South Africa.* Berkeley: University of California Press.

Gibbon, S., and S. Whyte. 2009. Introduction. Biomedical technology and health inequities in the global north and south. Special edition of *Anthropology & Medicine* 16, no. 2: 97–103.

Hardon, A., and H. Dilger. 2011. Introduction. Global AIDS medicines in East African health institutions. *Medical Anthropology* 30, no. 2: 136–57.

Heald, S. 2003. An absence of anthropology: Critical reflections on anthropology and AIDS policy and practice in Africa. In *Learning from HIV and AIDS,* ed. G. Ellison, M. Parker, and C. Campell, 210–38. Cambridge: Cambridge University Press.

Ingstad, B. 1990. The cultural construction of AIDS and its consequences for prevention in Botswana. *Medical Anthropology Quarterly* New Series 4, no. 1: 28–40.

Kleinman, A. 1999. Moral experience and ethical reflection: Can ethnography reconcile them? A quandary for 'the new bioethics'. *Daedalus* 128, no. 4: 69–97.

Mattes, D. 2011. 'We are just supposed to be quiet'. The production of adherence to antiretroviral treatment in urban Tanzania. *Medical Anthropology* 30, no. 2: 158–82.

Mattingly, C. 1994. The concept of therapeutic emplotment. *Social Science and Medicine* 38, no. 6: 811–22.

Meinert, L., H. Mogensen, and J. Twebaze. 2009. Tests for life chances: CD4 miracles and obstacles in Uganda. *Anthropology & Medicine* 16, no. 2: 195–209.

Mogensen, H. 2010. New hopes and new dilemmas: Disclosure and recognition in the time of antiretroviral treatment. In *Morality, hope and grief: Anthropologies and AIDS in Africa,* ed. H. Dilger and U. Luig, 61–79. New York: Berghahn.

Nguyen, V.K., C. Ako, P. Niamba, A. Sylla, and I. Tiendrébéogo. 2007. Adherence as therapeutic citizenship: Impact of the history of access to antiretroviral drugs on adherence to treatment. *AIDS* 21, suppl. 5: 31–35.

Rabinow, P. 1996. *Essays on the anthropology of reason.* Princeton, NJ: Princeton University Press.

Robins, S. 2006. From 'rights' to 'ritual': AIDS activism in South Africa. *American Anthropologist* 108, no. 2: 312–23.

Whyte, S. 1997. *Questioning misfortune. The pragmatics of uncertainty in Eastern Uganda.* Cambridge: Cambridge University Press.

Whyte, S., M. Whyte, and D. Kyaddondo. 2010. Health workers entangled: Confidentiality and certification. In *Morality, hope and grief: Anthropologies and AIDS in Africa*, ed. H. Dilger and U. Luig, 80–101. New York: Berghahn.

Wolf, A. 2001. AIDS, morality, and indigenous concepts of sexually transmitted diseases in Southern Africa. *Afrika Spectrum* 36, no. 1: 97–107.

'When there were only gods, then there was no disease, no need for doctors': forsaken deities and weakened bodies in the Indian Himalayas

Serena Bindi

Université de Nice Sophia Antipolis, Nice, France; Centre for Himalayan Studies (Centre d'Etudes Himalayennes), Villejuif, France

In this study the author analyzes the relationship between the individual body and the body politic in a region of the north Indian state of Uttarakhand, in connection with social changes occurring at the local and trans-local level, which are impacting the status of the different healing systems. By investigating these issues, this paper aims to shed light on some of the complex ways in which practitioners and patients who take part in a local method of healing, in this case ritual healing through possession, respond to the expansion of biomedicine.

Introduction

This paper is based on ethnographic fieldwork carried out between June 2006 and May 2007 in a network of rural communities in the state of Uttarakhand, in the central Indian Himalayas.

The fieldwork focused on the physical and interpretative itineraries set in motion by problematic events, with particular attention to healing practices based on possession rituals. The methodology consisted of participant observation of the everyday life of villagers; observation and filming of healing sessions; and informal and semi-structured interviews with healers and their clients about the problems that affect people's lives, about their healing choices, and about the outcomes of healing sessions.

The villages where the research was carried out, all located in the upper part of the Uttarkashi district, are inhabited by several caste groups ranging from Brahman to Rajput to Dalit sub-castes. The regional economy is based on subsistence agriculture, on remittances of those who have migrated to the Indian plains, on activities linked to the increasing presence of tourists and pilgrims in the area and, to a lesser extent, on animal herding and husbandry. In colonial times, this area was never subjected to the direct control of the British and remained a separate Hindu Kingdom under the name of Garhwal (Rawat 1989). Thus, the region was relatively isolated. Only in the 1960s, as a result of India's so-called 'Border War' with China in

1962, was the main trans-regional infrastructure built, while private as well as public development schemes followed in the wake of military use. In the last five decades, social and economic change has been swift and profound.

The research delineated a complex scenario of medical pluralism. As has been observed in many South Asian contexts (Amarsingham Rhodes 1984; Beals 1980; Lambert 1996; Nichter 1989; Pigg 1990; Pinto 2008), patients and sufferers continually resort to different healing techniques in order to find the best solutions to their problems. Alongside these different healing practices, biomedicine today occupies an important place in the treatment of illness. Biomedicine has been known in Garhwal for several decades, through the drugs that in the villages are administered by trained as well as unqualified practitioners (see also Pinto 2008), the presence of a hospital in the district's main city, the existence of Primary Health Centres all over the district, the flourishing of many private medical surgeries, and the periodic organization of government funded health-checking camps.

One striking outcome of the research was the way in which, during verbal exchanges within performances of possession as well as during interviews with healers and locals, people's relationship with biomedicine appears charged with ambiguity. While in practice people call regularly on all kinds of healers, in their theorizing about healing they focus increasingly on the competition between local deities and biomedical doctors. *Jyotiṣī* (astrologers), *vaidya* (learned practitioners of Ayurveda), *sādhu* ('holy men' who can provide healing through sacred verbal formulas, rituals, meditation practices and also herbal remedies), village health workers, nurse-midwives, vaccinators, veterinary assistants, tantra-mantra gurus (healers using tantric knowledge), *bākī* (diviners who undergo possession), experts in herbal remedies (*jarī būtī*) and practitioners of other forms of treatment (such as bone-setting, abdominal massage, cautery techniques) all fade into the background as people centre their discourses on biomedical doctors and local deities.

Within this local reflexivity on healing practices, the spread of biomedicine in Garhwal is at times spoken of as something that lessens the deities' power to heal and weakens people's bodies. As this paper will demonstrate, a correlation is made between the fact that deities are being forsaken by the people due to the success of doctors, and the weakening of people's bodies (their being subject to a greater number of illnesses).

This perception is well illustrated by the words of a Rajput man in the village of Jantu, located 30 km from the district capital, Uttarkashi. During an interview concerning the general health conditions of his village, the man suddenly stated:

> When there were only gods, then there was no disease, no need for doctors. Now there are so many illnesses. And for every illness there is need for a special doctor, there is need for a special medicine. Before, we didn't have all these illnesses.

Such words highlight the feeling of a disturbing social change where a weakened body politic is reflected in the feeling of disempowerment at the level of individual bodies.

In order to throw light on these matters, this paper takes into account two interrelated dimensions of the body that have been the focus of much anthropological theorizing about the body (Scheper-Hughes and Lock 1987): the 'body politic', referring to the political regulation, surveillance, and control of bodies (individual and collective), and the 'individual body', namely the individual understanding and experience of the body.[1] In this study the author analyzes the relationship between individual body and body politic in Garhwal in connection with social changes occurring at the local and trans-local level, which are impacting the status of the different healing systems. By investigating these issues, this paper aims to shed light on some of the complex ways in which practitioners and patients resorting to a local method of healing, in this case ritual healing through possession, respond to the expansion of biomedicine.

Healing and divination practices linked to the cult of territorial deities: a local body politic

While the medical landscape in Garhwal is, as elsewhere in rural India, highly pluralistic, there is an important factor that makes the Garhwali case different from medical pluralism described for other Indian and Nepali rural contexts (e.g. Beals 1980; Lambert 1996; Pigg 1990). This is the important role in people's everyday lives of practices of possession linked to the cult of territorial deities. Amongst the different categories of local deities who have an active 'social life' through being protagonists of institutionalized practices of possession, it is the village deities (Grām Devtā and Grām Devī) that, by way of healing and divination ritual practices, have the strongest impact on people's daily life. Each village's tutelary deity has a mobile form called a *dolī*, consisting of a wooden palanquin that can be carried on the shoulders of two ritual specialists. People regularly come to the village temple to seek the help of the god. In order for people to communicate with gods, two men – the deity mediums called *palgyār* – take the palanquin on their shoulders; at this point the deity enters their bodies, which in turn start to move. The two mediums' body movements make the palanquin itself swing in different directions. These movements of the palanquin are regarded as a special language through which the deity expresses its will, which can only be understood by an elite group of knowledgeable people who 'translate' it for the public. The entire institution is controlled by the villages' hegemonic groups, which in the past were linked to the central royal power and which belong to the two upper castes of Brahmans and Rajput. Both the interpreters and the mediums come from these social groups. In some villages, in addition to the palanquin the tutelary deity uses other forms of communication, all entailing possession, such as communication through the words of a medium (*paśvā*) or the sound of a ritual drum (*dhol*) played by a low caste musician. The legitimacy of the entire system is based on the belief that villagers are naturally subject to the will of the gods, who have control over the space of their village.[2] Local deities are considered kings over their territories, and the relationship that links them to their subjects is a combination of care and control. Virtually all personal or collective events must be brought before the village god's palanquin to be provided with meaning. Such events range from physical illness to unusual individual behaviours, emotional distress, household disagreements, subsistence practices, disputes between neighbours or relatives over land tenure, natural catastrophes, and sudden deaths. Aside from offering directions for action, territorial deities perform healing rituals that are often referred to as *rakhvālī*, which literally means 'protection' or 'the act of taking care of someone'.

The territorial deity's bodily technology

The power of care and control that characterizes the village deity's healing activity is played out within a specific body politic that has sometimes been referred to as 'government by the deity' (Sutherland 1998). Experiences of the body and its symptoms are continuously produced by living under the control of local territorial deities and being subject to their political regime. In other words, the territorial deities' body politic entails a bodily technology. This technology produces bodies that feel interconnected to the social body, to the village territory, to the soil and its productivity, to the territorial deity and its 'happiness', and that feel strengthened when submitting to the local deities' performances of diagnosis and healing.

Local deities' body technology involves, for example, the activities of carrying the heavy wooden palanquin on one's own shoulders (as in the case of the two *palgyār* or porters); of making it dance during village festivals (an action that most men belonging to the higher castes have to perform); of accompanying the deity on its pilgrimages, sometimes barefoot; of undergoing long and physically exhausting possessions (as in the case of deities' mediums called *paśvā*); of enduring painful flagellations (as during the séances of exorcism executed by local deities); of bending down under the cloths (*patolā*) of the palanquin for long periods while the deity confers blessings (*āśīrvād*) and ritual formulas (*mantra*) on the sufferer or on some healing substances; of meeting certain requirements of bodily cleanliness at specific times; of following dietary rules, namely the avoidance of alcohol, meat and restaurant food (rules that have to be followed by the deities' mediums and by devotes asking the deities for relief from suffering). As an integral part of the individual's submission to the power of territorial deities, these activities contribute to the fact that, at the end of the healing process, the person often feels a sense of strengthening the body.

The three cases that follow provide ethnographic evidence for these observations. On the one hand, these examples highlight the way in which devotion to territorial gods implies certain bodily activities, which are sometimes especially physically taxing. On the other, they highlight how, at the end of the healing process, people often affirm that they feel a sense of health or bodily empowerment.

The first example concerns a 14-year-old boy from a Rajput family in the village of Dharali. Too busy in working in their fields, his parents had decided not to accompany the territorial god on his annual pilgrimage to Gangotri but to send their son in their place. During the night, the boy started to suffer 'a lot of pain in one leg, and the leg became very swollen'.[3] After interrogating the palanquin of the territorial god, the reason given for the pain was that his parents had refused to follow the deity on the annual pilgrimage. The god told them to make a votive offering (*uṭhānū*) and to promise to accompany the deity on his pilgrimage if the child recovered. The morning after, the child felt better: 'the leg was exactly as if nothing had happened to me'. The case well exemplifies how devotion to territorial deities is strictly related to bodily activities; in this case it involves the performance of a pilgrimage.

Another case concerns a Brahman family living in Baghial village, near by the district main town of Uttarkashi. The elder son of the family, Sunil, was regularly possessed by the deity of Baghial village. A few months after Sunil had left the village to join the navy, his father experienced severe pain in the lower part of the belly. Urgently hospitalized in the main district clinic, he was operated on immediately for appendicitis. After the operation, which was declared successful, he went home but failed to recover from the pain in the belly and all over the body. When he went back to the hospital to have the stitches removed, doctors found that everything was satisfactory. Back in the village, he consulted the god, who told him that the only way to get well was to make his son return. The punishment (*dos*) of the god had fallen on the family. After Sunil left the navy and returned to the village, where he started again to undergo possession, his father felt completely cured. As above, the case of Sunil's father shows how, within this body politic, someone's body may be in pain due to a member of his social group disrupting the harmony with the territorial deities. Likewise, a feeling of individual bodily strength can emerge from the fact that one's relative (in this case the man's son) surrenders devotionally to the deity's will.

In the multi-caste village of Pata, a 16-year-old girl from a Rajput family, suffering from stomach pain, episodic fainting and aggressive behaviour was sent by the village god medium to see a doctor. At first her family bought some ayurvedic medicine from the market (a herbal tonic for 'strength'), but later when the symptoms persisted, she went through several admissions to the district hospital. Diagnosed with severe gastritis, she was treated with a herbal calming tonic, antibiotic medication (the doctors told the author that it was to kill any H. pylori present) and acid blockers. Nevertheless, each time after returning home she started complaining again of stomach pain and behaving aggressively. When the family decided to consult the village god once again, he ordered them to organize an exorcism to get rid of the ghost of a woman who had died while giving birth in a nearby village. During a fairly violent ritual, consisting of several phases all orchestrated by the village god's palanquin, the ghost was invited to manifest itself in the girl's body and talk; then the girl (or rather the ghost inside her) was beaten by several mediums, and she was finally inducted to vomit before falling unconscious to the floor. At the end of the ritual, after receiving the territorial god's *mantras* and a protective amulet, she told the anthropologist how the hospital had been a much less painful and more pleasant experience. She said she was now feeling pain all over her body and that the next day her face would be swollen and ugly due to the beating she had received. Nevertheless, she added, 'Tonight I will be sleeping between my parents because I am still very scared, but from tomorrow I will feel fine!' In this case, enduring flagellation is part of the subject's moral and devotional duty, as well as a necessary path to healing. Moreover, the sufferer affirms that she feels an empowerment of the body.

Taking these cases together, one notices how local healing is not always the first resort (see also Lambert 1996). On the contrary, when possible and available (for example in villages like Pata and Baghial, which are located close to the road leading to the district's main city), biomedicine is either the first option or a therapy which is being pursued alongside ritual healing. Nevertheless, as was also shown by Amarsingham Rhodes (1984) in a Sinhalese context, even when the efficacy in biomedical terms could be attributed to biomedicine, people often attribute it to the territorial deity. As Amarsingham Rhodes points out, this is not only due to the fact that the diagnoses given by deities are powerful polyvalent symbols that can confer meaning and sense on individual experiences in religious and social idioms. It is also due to the fact that the patients undergo a powerful performative aesthetic experience, which makes them feel healed (Desjarlais 1996). The performative practices linked to the institution of the territorial deity – with the ritual traversing of its space, the rich aesthetic of dancing and oscillating in different directions, and the variety of bodily activities involved in its healing and divination sessions – make possible the experience of bodily empowerment, together with the feeling of the link between the body, the deity, the territory and the community.

Neglected gods, weakened bodies

In many instances, therefore, while the patient undergoes different treatments, the ultimate healing efficacy is attributed to territorial gods. Nevertheless, as the cases above demonstrate, it is evident that biomedicine today occupies an important position in people's health-seeking strategies. In addition to its acclaimed efficacy, especially in treating certain kinds of diseases, the popularity of biomedicine is linked to the fact that it is the main form of Indian Government health service. The growing popularity of biomedical treatments may also be explained by the fact that their faster efficacy is more suited to the ideologies of productivity and speed, which are spreading at a time of dramatic economic and social change (Halliburton 2009). It is also important to stress that biomedicine enjoys, in the hills of Garhwal as in other local contexts, a certain amount of prestige value owing to its associations with modernity, the urban lifestyle and cosmopolitanism (see also Lambert 1996; Pigg 1990). However, the variable discourses regarding medical pluralism and biomedicine also show at times no coherence with people's actual behaviour.

At times, it would seem that biomedicine has been quite unproblematically integrated into the local aetiological system and into people's health-seeking itineraries. In some conversations with locals, gods and doctors and their systems of healing seemed to complement one another. While illnesses (*rog*) that have a purely physical cause (*bhautik*) and stem from physiological processes can only be treated by doctors, deities' possessed mediums are the only experts in curing forms of sorrow deriving from the evil eye (*nazar*), distress originating from attack by external spirits (*bāharī ātmā*), sorrows deriving from the falling on some person of the shadow (*chāyā*) of fairies, problems caused by the action of a deity initiated by a person against his enemy (*ghāt*), and troubles triggered by the rage of a god when offended by personal or collective human behaviour (*doṣ*). According to this discourse, doctors and gods seem to be granted different fields of competence depending on the origin of the illness, whether natural or supernatural. This discourse legitimizes the ongoing healing power of local gods despite the spread of biomedicine.

On other occasions, the relationship between ritual healing and biomedicine is seen as more problematic. The healing power (*śakti*) of gods is sometimes spoken about in nostalgic terms, as in the sentence that gives this paper its title, and the blame for the gods' diminished power to heal is put on the spread of biomedicine. This argument is often adopted by both priests and lay persons to explain and justify ritual failures, as for example when an exorcism or a healing ceremony does not prove efficacious even after a long time and several attempts. People often explain the persistence of the problem with a discourse that entails a circular feedback loop. Increased recourse to 'modern' ways of healing, and increased adoption of a modern lifestyle generally, diminishes people's faith in gods and induces devotees to forget their ritual duties towards them. This in turn lessens the local deities' power to heal, a power that is perceived as based on a relationship of reciprocal exchange between belief (*śraddhā*) and celebration (*pūjā*) on the people's side and the bestowing of bliss and healing by the gods. The weakening of the deities' power to heal induces more recurrent ritual failures, which in turn increases demand for biomedical remedies. In this reasoning, modernity, in the form of biomedical power, is thought to have broken a rule of reciprocity on which the moral economy of ritual healing is based. This rule, which has been widely recorded in other Hindu contexts, governs both spiritual and human relationships; the superior being (for example a god or a human being of higher status) receives veneration and respect, and gives bliss and help in return.

A local priest from Deepa's village commented as follows on the healing functions of his village deity:

> Kandar Devtā, Someshwar Devtā, Naag Devtā these are all local gods because they are accepted and worshipped by local people. As long as you worship them, they will help you. In Gita [Bhagavad Gita] it is said: 'To us who accept and worship, he only will help us . . .' Now people believe less and lose faith in gods, they worship doctors. Gods get angry as they don't get celebrations and offerings. These gods will not help you. Therefore today people are not getting any benefit. If people don't show feeling and respect for the god the god becomes less effective. This means that as long as we have respect for the god, things will go smoothly, but the moment we become careless toward the god, he stops helping us.

A second argument that is employed to talk about the problematic relationship between healing and biomedicine in Garhwal centres on the issue of bodily discipline. After the territorial deity had tried for a long time to cure the lung of a 35-year-old man who, nevertheless, failed to recover and was hospitalized in the district clinic, a temple priest observed:

> Our bodies are becoming weak. People are not able to refrain from smoking, from consuming alcohol. They are not attentive to what they eat. So then they need doctors. Gods will keep helping us only if we have good thinking, if we do good work, and if we keep our bodies strong and healthy. Now people eat things which are not clean, they eat animals, they drink alcohol and if they don't pay attention to the things they do with their bodies, then the gods will run away far from these people and their bodies will become more and more weak. Then people eat so many pills when they are sick and need to recover fast. In the past, we didn't need all these pills.

According to the priest, the sufferer – like so many other patients who come to see the local gods – does not take proper care of his body. It has been argued that, in Hindu contexts, a correlation is made between one's bodily state, bodily discipline, and the moral and ritual condition of the person, in particular the right of the person to enter into relations with deities (e.g. Parry 1989). The words quoted above indicate a vicious circle: the loss of self-discipline by people in Garhwal prevents them from being in the state of ritual purity that is needed to perform worship and maintain a close relationship with deities, and, as a result, the gods no longer support or heal them. This, in turn, will contribute to the weakening of people's bodies, enhancing their need for constant repair by the biomedical system. The body is here taken both as the mirror of, and the possible agent for, a reform of this state of affairs.

Discussion

Local reflexivity about the body politic and especially about the impact of biomedicine upon ritual healing is imbued with a certain ambivalence. It is here argued that an understanding of these different discourses that are emerging in contemporary Garhwal can be achieved by taking into account several interrelated dimensions, namely the local body politic (involving deities controlling and protecting the territory and local bodies), the individual bodily experience, and the impact of contemporary social change on the status of biomedicine as well as on local practices and perceptions of the body.

First, on a political level, one can see how these discourses might be motivated by a fear of the elites controlling the workings of the palanquin, an important means of control that is part of everyday village politics. The prestigious status of biomedicine can be perceived as eroding the political power of local deities and making their verdicts less compelling.

However, one might also see in these discourses the penetration of trans-local discourses about healing into the local setting, a way of reflecting about social change and a social means to position oneself vis-à-vis this perceived change. It is remarkable how the wide range of practitioners available in the area fades into the background as people focus predominantly, in theorizing about healing, on the potential compatibility or competition between deities and doctors. National and trans-local discourses about progress in many contexts have made biomedicine into a symbol of modernization and cosmopolitanism (see also Pigg 1990), as opposed to the hill dwellers' superstitious practice of ritual healing through possession. The dichotomy has entered local ways of thinking and introduced a new kind of reflexivity. To stress one's disbelief in healing through possession might become a means of integration into this 'modern' society. To underline the higher power of one's own territorial god may, in turn, serve to defend one's identity and to state the possibility of being modern while at the same time preserving one's own traditions.

Finally, at the level of the individual body, this paper has demonstrated that being subjected to the power of territorial gods, their diagnoses and their healing sessions involves bodily activities that, taken together, can bring about a person's renewed sense of health and make them feel, in different ways, the interconnectedness between their individual body, the collective body and the territorial deity.

Nevertheless, nowadays Garhwalis are not only local ritual subjects. They are at the same time mobile actors, living in an interconnected world and exposed to trans-local discourses about modernity and progress. Consequently people's embodied sense of self is simultaneously shaped by different local practices and discourses. Now, more and more locals start to feel that it is not only more efficacious but also more prestigious to undergo biomedical treatment, vis-à-vis their integration in the larger national context. Conversely, the experience of the body as imbued with power given by the territorial god and dependent on the moral conduct of all community members is increasingly linked, in people's everyday perception, with the stereotype of their marginality and supposed backwardness in the eyes of the wider society. Ritual healing becomes at times embarrassing or inappropriate in the changing social field. But this practice, with the bodily experience that it produces, is nevertheless part of people's everyday experience. A form of 'hysteresis' (Bourdieu 1972) may be produced in people's lives by a discourse that confers on ritual healing, namely their everyday experience, a status of superstition. This perception may contribute to the emergence of the critical discourses that have been analyzed here, which focus on the neglect of deities by the people and the consequent weakening of their bodies. These discourses show that, while making use of biomedicine, social actors perceive and express in religious and somatic terms the destabilizing effect that the encounter with biomedicine is producing, not only on local politics but also on their bodily knowledge and experience.

Acknowledgements

The author thanks Elisabeth Hsu, Caroline Potter, Federica Fratagnoli and three anonymous reviewers for their helpful suggestions on previous drafts of this paper. Funding for this research was generously provided by a fellowship of the University of Siena and by a grant of the French-Italian University (Université Franco Italienne UIF). The author followed the ethical guidelines of the European Association of Social Anthropologists.

The text is based on a paper given at a conference funded by the Royal Anthropological Institute, the Wellcome Trust and the Institute of Social and Cultural Anthropology, University of Oxford.

Conflict of interest: none.

Notes

1. Scheper-Hughes and Lock mention also a third kind of body, the social body, namely the body as a symbol or map with which people shape or understand their environment (Scheper-Hughes and Lock 1987).
2. This complex system of divine jurisdictions is found in several nearby areas of the central Himalayan region. See also Berti (2001), Bindi (2009), Sax (2002: 157–85), Sutherland (1998), Vidal (1988).
3. Interview with Sajal Panwar, Dharali, 09-2006.

References

Amarsingham Rhodes, L. 1984. Time and the process of diagnosis in Sinhalese ritual treatment. *Contributions to Asia Studies* 18: 46–59.

Beals, A. 1980. Strategies of resort to curers in south India. In *Asian Medical Systems*, ed. C. Leslie, 194–5. Berkeley: University of California Press.

Berti, D. 2001. *La parole des dieux. Rituels de possession en Himalaya indien.* Paris: CNRS Éditions.

Bindi, S. 2009. La fabrique de l'événement. Paysage de possession, de soin et de production de sens dans la vallée du Baghirati (Garhwal, Inde du Nord). Thèse de Doctorat, Università di Siena and EHESS, Paris.

Bourdieu, P. 1972. *Esquisse d'une théorie de la pratique, précédé de trois études d'ethnologie kabyle.* Paris: Seuil.

Desjarlais, R. 1996. Presence. In *The performance of healing*, ed. C. Laderman and M. Roseman, 143–64. London: Routledge.

Halliburton, M. 2009. *Mudpacks and prozac. Experiencing Ayurvedic, biomedical and religious healing.* Walnut Creek: Left Coast Press.

Lambert, H. 1996. Popular therapeutics and medical preferences in rural north India. *The Lancet* 348, no. 21/28: 1706–9.

Nichter, M. 1989. *Anthropology and international health: South Asian case studies.* Kluwer: Dordrecht.

Parry, J. 1989. The end of the body. In *Fragments for a history of the human body*, ed. M. Feher, R. Naddaff, and N. Tazi, Part Two, 491–517. New York: Zone.

Pigg, S.L. 1990. Disenchanting shamans: Representations of modernity and the transformation of healing in Nepal. Unpublished PhD Thesis, Cornell University.

Pinto, S. 2008. *Where there is no midwife. Birth and loss in rural India*. New York: Berghahn Books.

Rawat, A.S. 1989. *History of Garhwal, 1358–1947*. New Delhi: Indus Publishing Company.

Sax, W.S. 2002. *Dancing the self. Personhood, performance and the Pandav Lila of Garhwal*. Oxford: Oxford University Press.

Scheper-Hughes, N., and M. Lock. 1987. The mindful body: A prolegomenon to future research in medical anthropology. *Medical Anthropology Quarterly* 1, no. 1: 6–41.

Sutherland, P. 1998. Travelling gods and government by deity: An ethnohistory of power, representation and agency in west Himalayan polity. Unpublished PhD Thesis, Oxford University.

Vidal, Denis. 1988. Le culte des divinités dans une région d'Himachal Pradesh. Coll. Études et thèses. Paris: Éditions de l'Orstom.

Second nature: on Gramsci's anthropology

Giovanni Pizza

Dipartimento Uomo & Territorio, University of Perugia, Italy

The aim of this paper is to convey the relevance of a Gramscian perspective in medical anthropology, stressing his anti-essentialist way of reasoning about 'nature'. The author claims that Gramsci's understandings of the bodily life of the state can deconstruct naturalized realities in ways that are helpful for the ethnographer engaged in the political anthropology of embodiment and the management of health, persons, and life itself. The paper is presented in three parts. An attempt is made, first, to frame the relevance of Gramsci for Italian medical anthropology and second, to explore the components of the Gramscian concept of 'second nature' within the perspective that he himself calls 'an anthropology'. Third, an example is given of how the proposed Gramscian insights could inform an ethnography on the biopolitical aspects for the early detection of Alzheimer's disease, which is currently being carried out in Perugia.

The relevance of Gramsci for medical anthropology

Antonio Gramsci was born in Ales, Sardegna in 1891 and died in 1937, due to the serious consequences of his 11-year imprisonment by Italian fascists. During his lifetime, his works were known only to militants of the international communist movement and the Italian Communist Party, which he helped constitute in 1921 (Fiori 1990; Santucci 2010). Although the agony of imprisonment aggravated irremediably his already precarious state of health, he continued his political praxis through his daily writing. Besides his correspondence, he dedicated himself to writing notebooks (from 1929 to 1935) which began to be published only in 1948, three years after the defeat of the fascist dictatorship and 11 years after his death. Therefore, his works began to influence critical thought in all disciplines only from the 1950s onwards. Today, Gramsci is the internationally most well-known Italian intellectual. His works are divided into three blocks of writing: *Prison notebooks*, *Pre-prison writings* and *Letters from prison*.

Gramsci's oeuvre greatly influenced Italian anthropology from the 1950s to the 1970s. His observations on folklore were considered a legacy for the study of subaltern cultures (Cirese 1976) and produced a new phase of anthropology through the work of Ernesto De Martino (1908–1965), whose writings set in motion a long-lasting debate over the changes taking place in popular culture in the new democratic

Italy (De Martino 1949, 2005 [1961]). In medical anthropology the influence of Gramsci led, in the second half of the twentieth century, to studies on the relationship between 'official' and 'popular' medicine (see Seppilli, this issue). We have also seen a return to Gramsci in the international field. Ronald Frankenberg (1988) was the first to exhort an attentive, direct reading of Gramsci for medical anthropology in the English-speaking world. More recently, Kate Crehan (2002) has demonstrated how Gramsci elaborated an anti-essentialist notion of culture, considering it as a dialectical process involving knowledge and power, within a field of historical forces.

The author's current re-reading of Gramsci is more concerned with understanding his anti-essentialistic approach in reasoning about 'nature'. In the author's reading, Gramsci furnishes useful tools for the ethnographic exploration of two directly connected spheres – the sphere of life and that of politics – and thus prompts work in the arena of 'biopolitics', a term launched by Foucault to define this nexus. What has emerged from a long-term Gramscian seminar in Perugia is not so much the importance of an 'anthropological use' of his theories, but his outright *anthropological vocation*. Critical readings of his work offer interpretations on the question of the relationship between the body and the state, which anticipate other great masters of critical thought like Michel Foucault (1926–1984) and Pierre Bourdieu (1930–2002) – authors who came a generation after Gramsci, yet who have been more influential in medical anthropology (Samuelsen and Steffen 2004).

This is not the place for a comparison between Gramscian critical thought and theirs. Gramscian debate has not neglected this comparison.[1] The following section will instead demonstrate the bodily dimension of hegemonic processes on which Gramsci reflected in prison. Unlike Foucault or Bourdieu, Gramsci is not interested in the foundation of a new critical theory of social reality that can then provide the tools for praxis. He instead elaborates a living theory that reflects its constitution in the concrete experiences of real life. These experiences, in turn, could affect the social environment by giving rise to an initiative of will, a dialogue of transformation, a dialectic of hegemony. The complexity of the concept of hegemony implies therefore the capacity to interpret it as a plurality of embodied practices that are carried out in a framework of relations of force, that are at times stable and at times in flux, but in any case never fixed or final. This interpretation frees Gramsci from the old Marxist debates (which have in some ways hampered interpretations of Foucault and Bourdieu, by concealing the bodily aspects). It also allows his work to have a direct relevance for anthropology that is less theoretical and more political, because it is able to account for the power relations not only on the side of dominion but also on that of the dominated, whose subalternity is nonetheless temporary and reversible. In order to understand the bodily forms of such a dialectic, Gramsci's 'second nature' will now be explored.

Writing (against) nature

In *Prison notebooks*, in a Note entitled 'Arguments of culture. Against Nature, Natural...', Gramsci criticizes the idea of 'human nature' and expressions like 'natural' or 'second nature'. Deconstructing human nature means understanding it as the whole complex of social relations. In writing *against nature* he is not simply advocating for culture and historicity of human beings. Rather, he is underlining the habitual, embodied dimension of nature and reality.

> What does it mean to say that a certain action is 'natural' or, that it is on the other hand 'against nature'? Each of us, deep within ourselves, believes that we know exactly what this means, but if we ask ourselves a direct question, we realise that it is not so simple to answer. In any case, we need to realise that we cannot talk of 'nature' as something which is fixed or objective; in this case 'natural' means right or normal according to our current historical conscience, which in turn is our 'nature'. The nature of humankind is a set of social relationships which determine an historically defined conscience, and it is this conscience which indicates what is 'natural' or not [and thus we have a human nature which is contradictory because it is the set of social relationships]. (Gramsci 1975 [1948–1951], 1032) [The author's translation]

> That 'human nature' is the 'complex of social relations' is the most satisfactory answer, because it includes the idea of becoming (man 'becomes', he changes continuously with the changing of social relations) and because it denies 'man in general' (Gramsci 1975 [1948–1951], 885 [SPN, 355]).

> We speak of 'second nature'; a certain habit has become second nature; but was 'first nature' really 'first'? Is there not in this common sense expression the hint of the historicity of human nature? (Gramsci 1975 [1948–1951], 1032) [The author's translation]

The concept of 'second nature' is at the heart of the Gramscian theory of hegemony, to the extent that it identifies the bodily dimension of this complicated notion, a hidden dimension in the classical Marxist debate. It involves a phenomenological and political attention for the 'living' that has not always been well understood.[2] According to Gramsci, 'second nature' is the internalization of what he calls *abitudini di ordine*, 'habits of order' (Gramsci 1975 [1948–1951], 138, 2160 [SPN, 298]).[3] As hegemony is not to be understood only with respect to domination but also in regard of possible change, second nature is likewise always subject to possible transformation – one can rebel against it, in a self-reflexive process of dis-embodiment, with the creation of a new order in sight, and therefore of a new second nature. Hegemony is therefore a complex concept, which includes agency and the transformation of persons, three concepts explored more closely in what follows.

Hegemony has been the Gramscian concept that has received most attention, but it is also the most misunderstood.[4] Sometimes it is interpreted solely as the function of domination exercised in the framework of opposition between the 'hegemonic' and 'subaltern' classes. This dichotomy is misleading. Those who read Gramsci will not find in his thought a separation between hegemonic and subjugated culture, but an underlining of the intimate dimensions of the hegemonic dialectic as observed in the relationship between the body and the state. The state is considered an 'active and permanently active centre of its own culture' (Gramsci 1975 [1948–1951], 1872 [FSPN, 68]). In Notebook 22, *Americanismo e fordismo*, Gramscian attention for the state is grounded in the centrality of workers' bodies, of their exposure to transformation in the process of production. The state functions as a 'body factory'. But hegemony is not only the state power exercising its authority through coercive action (including the organization of a naturalized 'spontaneous' consent by its subjects). Hegemony is also the politics of transformation, exercised through the critical capacity to denaturalise one's own body, thus bringing to light the dialectical interaction between the state and the intimacy of subjects.

Therefore the question of agency is crucial. The hegemonic relationship is dialectical because it is also active on the side of the critical agency of the subjects acted upon by the state. This is why, starting from the centrality of the working body, Gramsci (1975 [1948–1951]) reviews various fields of bodily experience that are acted upon by the 'permanent cultural activity' of the state; he identifies these in 'sexual obsession', in the construction of the female body and personality, and in the family. These are strategic fields in which one is acted upon but where, at the same time, one can also act with a view toward transformation. Taylor's cynical expression labelling the factory worker as a 'trained gorilla' is dismantled by Gramsci to show how it can be ironically overturned by the agency of the worker, who can orient the physical transformation to which he is subjected towards non-conformist actions. What Gramsci calls the 'animality' of bodies cannot be easily domesticated. This 'animality' has no 'psychoanalytic' tone. It is *una pratica reale 'animalesca'*, a real 'animal' practice, which prevents physical bodies from effectively acquiring the new attitudes (Gramsci 1975 [1948–1951], 139, 2163 [SPN, 300]).

Gramsci's constant attention to processes of subjectivization produced in the hegemonic dialectic leads him to a pioneering disarticulation of the *category of person*, which is relevant for anthropology. Those who have read Gramsci might have been struck by a new word that he uses quite often, both in his letters (with reference to himself) and in his analysis of political and cultural criticism (when referring to the state): the term *molecolare*, 'molecular'. This term is evidence of the experimental character of Gramsci's writings. Personally engaged in working class struggles and searching for new expressive forms that emerge from that transformative experience, Gramsci looks at the term 'molecular' for the possibility it offers of referring to the minimum unit of life experience, to the immediate detail drawn from daily life. It is a word-bridge between politics and the body. The notion of 'molecular' is used by Gramsci to describe the transformation of both the state and the person. In the last part of the Notebooks, Gramsci evokes the apologue of the shipwreck survivor recounted by Edgar Allan Poe in *The narrative of Arthur Gordon Pym*. In conditions of extreme hardship following a shipwreck, some men who would have sworn they would kill themselves first, eventually resort to cannibalism. But are they really the same persons? Gramsci's answer is No, because between the two moments there has been, due to the force of necessity, a process of 'molecular' transformation (Gramsci 1975 [1948–1951], 1762–1764 [FSPN, 108–111]). In a letter written in those same days to his sister-in-law (Gramsci 1994 [1947], 278–279), Gramsci recalls the same example, where the metaphoric force of shipwreck is now revealed as the real terroristic force of the fascist state prison on his own body and person. At the same time, he emphasizes the fluidity of the transformative process that unfolds in a manner that is totally embodied, intimate, and inarticulable. His reflections and accounts of his own illness interweave lived experience and the critique of the state, in an unceasing process of self-objectification. The body and his sufferings are considered as the field of a struggle of forces.

These insights define the outline of what the author has called Gramsci's *anthropological vocation*. Indeed, it is Gramsci who self-defined his philosophy of praxis (a term he used to refer to historical materialism) as a 'living philology' and, in another section of the Notebooks, as an 'anthropology':

> One may say that the economic factor [...] is only one of the many ways in which the
> more far reaching historical process is presented (factors of race, religion etc.), but it is

this farther reaching process that the philosophy of praxis wishes to explain and exactly on this score it is a philosophy, an 'anthropology', and not a simple canon of historical research. (Gramsci 1975 [1948–1951], 1917 [FSPN, 424])

Such an 'anthropology' allows him to broaden his vision of the relationships between the state and the body. Power is no longer considered a unique force, abstract and external, but is observed in its molecular fragmentation. The state is reconsidered as 'the entire complex of practical and theoretical activities with which the ruling class not only justifies and maintains its dominance, but manages to win the active consent of those over whom it rules' (Gramsci 1975 [1948–1951], 1765 [SPN, 244]). For Gramsci the state takes on the task of elaborating *un nuovo tipo umano*, 'a new human type'[5] (Gramsci 1975 [1948–1951], 2146). It acts, therefore, in a mutual intimate dialogue with its citizens. Gramsci is suggesting that if '"State" means especially conscious direction of the great national multitude, [then] a sentimental and ideological "contact" with such multitudes is necessary' (Gramsci 1975 [1948–1951], 1122, 2197, the author's translation). This formulation has led to a greater stress on material, practical, and thus political aspects of embodiment, and to a call for studying power in its concrete manifestations, as in the practical life of institutions. Following Gramsci's suggestions, we can consider the state as enlarged, fragmented and living in the intimacy of daily life. Attention to the molecular dimension of the state, and to the bodily production of common sense that legitimates it within civil society, lies at the heart of such critical work. Gramsci overcomes the obstacle, which derives from the split between state and civil society, by collapsing this dichotomy and making it part of the complex overlapping of molecular state powers. In Gramsci we find a *molecular anthropology*, the micro-physical examination of the processes that impact upon the dialectical relationship between the body and the state. The expression 'molecular' is a way of capturing with language the embodiment processes by positioning oneself as closely as possible to bodily experience.

The ethnographic usefulness of Gramsci's insights on second nature are demonstrated in the following section, by focusing on the mingling of state and body in the concrete habituated practices of the detection process for Alzheimer's disease.

Ethnographying Alzheimer's as second nature

In January 2009, new ethnographic research began, in collaboration with the Memory Disorders Centre, Hospital 'S. Maria della Misericordia', Department of Neurology (University of Perugia), entitled *Early diagnosis of cognitive deterioration with the aim of improving 'person-based treatment' – an integrated approach between general medicine, clinical neurology and medical anthropology*.[6] The project was promoted by the Umbrian Region and reflects its current health guidelines, which aim to establish policy that is more focused on the 'person' and his or her network of relationships ('social capital'). Thus, anthropologists have been acting in the policy-making and clinical-health-university fields, within a complex crossover of hege-monic dialectics made evident by the polyvalent dimension of Alzheimer's disease. Being a chronic degenerative pathology that transforms the 'person', Alzheimer's constitutes a prime example of the complex plurality of molecular practices that converge in the construction of a second nature[7].

In order to better understand the different aspects of our ethnographic participation, it is important to realise how this participation has effectively modified this network of relationships from the very beginning. At their first meeting with the neurologists in 2009, the anthropologists realised that the main task that they were being asked to carry out was to help the neurologists to convince the general practitioners to administer a new questionnaire to patients, in order to detect signs of cognitive disorder. The project constituted the second phase of an already consolidated experimentation, which began in 2006 and which was carried out by the neurologists from the University of Perugia who had formulated, together with a private university in Rome, the *Basic Italian Cognitive Questionnaire* (BICQ), a technical tool that was considered fundamental for an early diagnosis of cognitive deterioration. For the biologists, the BICQ constituted the core of the project. The general practitioners were to be divided into two large groups – one group would administer the BICQ questionnaire to all their patients over age 60, while the other group would continue to rely solely on their own medical experience. Comparative research should verify its diagnostic usefulness, devised by the neurologists on the basis of its 'simple' and 'speedy' effectiveness:

> The Bicq is designed as a routine instrument for screening patients with initial cognitive impairment in daily practice. It is devoted to the general physician who needs a tool for deciding whether a subject deserves further diagnostic investigation in specialised centres. The administration of Bicq is *easy, fast and does not require any training, since it is composed of 12 simple and ecologic questions referring to daily life.* (Giaquinto and Parnetti 2006, 123; author's emphasis)

Ease, speed and training-free are the qualities extolled in this new tool, which is projected to improve collaboration between general practitioners and clinical neurologists. Its design is founded on basic assumptions of everyday life, and it aims to account for the 'cultural' through stereotypes of the presumed daily life of 'Italian elderly people' or of 'Italian food culture' (bread, pasta). The BICQ consists of 12 questions. If the patient obtains a score greater than nine, he or she is healthy; if lower then he or she is considered 'at risk' and referred to the Centre for further diagnostic analysis. The test questions reflect four areas of cognitive evaluation:

> **Personal orientation**: 1. What is your age?/ 2. What is your date of birth?/ 3. Where were you born?/ 4. Until what age did you go to school?/ 5. Where do you live? – **Reality orientation**: 6. What day is it today?/ 7. Who is (are) the person(s) with you? – **Family**: 8. What are your parents' names?/ 9. What are your grandchildren's names? – **Shopping**: 10. Can you indicate the price of bread per kilo?/ 11. Can you tell me two brands of pasta?/ 12. What change do you expect to have in return from a banknote of Euro 10.00 if you have to pay Euro 6.50?

It is not necessary here to deconstruct the obvious cultural reductionism of the questionnaire. In the Gramscian anthropological perspective, even if the textual dimension and its practical application cannot be separated, it is more important here to underline the emergence of contradictions in the practical application, rather than to stigmatise stereotypes and conceptual dichotomies present in the text.

According to the neurologists, the anthropologists had the role of mediator, in whose ethnographic experience, however, the communication seemed to be compli- cated precisely by their device. Rather than being a simple tool capable of detecting 'persons at risk', the BICQ served to unveil the composite, conflicting and polemological nature of the collaborators. The anthropologists' participation in the project led to the identification of the early diagnosis of Alzheimer's as a set of molecular practices, which had a declared aim of fabricating a specific form of second nature, namely the bio-pathologisation of senile age. This was achieved by means of a diagnostic codification of a transitory state, 'Mild cognitive impairment',[8] that constitutes the precursor to Alzheimer's disease – or rather, the social and laboratorial sphere in which it coagulates. Such a second nature, however, can only be produced through a hegemonic process, which can have unforeseeable results. By highlighting that the BICQ was in practice much more complicated and not at all 'quick and easy', the ethnographers in fact ratified its failure with respect to those objectives outlined in the scientific rhetoric. The ethnographers also demonstrated that the BICQ could activate unexpected regimes of interaction. Some examples included alliances between general practitioners and patients in contesting the price of bread; shared ironic comments on the alienation contained in the question 'Who is the person with you?', when in reality the only people present were the doctor and the patient; the emergence of strong emotions in patients who remembered their lost parents, evoked by the questions concerning family; and feelings of rebellion or anxiety aroused by answering questions that juxtaposed disease, family relations, intimate memories and commonplace daily activities.

The denial of the 'simplicity', put forward as a reason for using the BICQ, reveals a contradictory context in which both the simplification of the questionnaire as well as its failure serve to manipulate the power relations that led to the fabrication of Alzheimer's disease as a second nature. The ways in which 'Mild cognitive impairment' and consequent Alzheimer's disease are *done* or *undone* (Mol 2002) are therefore fluid, and depend on the actual relationships constructed through a multiplicity of actions, and through critical reflection upon them. Second nature is always at stake, and it confers upon the process an uncertainty, which is tied to different variables, such as the degree of reflexivity and presence during the medical visits, the patients' dialectics of rebellion and agency, and the institutional applications of the research and their bureaucratic component. Even if the contradictions that the ethnography highlights appear to be evident – (im)possible sick/not sick patient, based on the score obtained from the questionnaire – in the beginning, they become molecularly articulated in a more and more refined way in subsequent phases in which the patient, who does not obtain a sufficient BICQ score, finds himself or herself.

There are multiple stages of diagnostic analysis, which are dislocated in time and space and disseminated in specific clinical contexts. The patient who obtains a score of less than nine is directed to the Centre, where there are six attending doctors specialised in neurology who are directed by a university researcher, and a psychologist who administers psychometric tests. The patients undergo a further series of tests, some on the spot and others by appointment. In the case of a positive score, further diagnostic examinations are carried out, which might be technologically invasive (such as rhachiocentisis). These too spread out over time due to 'waiting lists' and are delegated to other hospital specialists. The search for rapid results in early diagnosis contrasts with the lengthy procedure times. On the other hand, the rhetoric of time is present not only in the clinical procedure, but also in the linguistic exchanges between doctors and patients on the question of ageing. Even as some memory disorders are thought to be due to the 'natural' process of ageing (to which the patient is often advised 'to get used to' and 'accept'), the dialectic of second nature nonetheless comes into play in constructing a pathology of ageing, also for those who are considered 'within the norm for age and education' (as the diagnosis of 'normality' states in these cases).

The question of time pervades, therefore, the whole course of the diagnostic process, manifesting itself in quite solid and corpulent forms. From the very beginning it becomes a material ingredient, which is indispensable in the clinical construction of body-political reality. The preliminary meetings between the doctors, neurologists and anthropologists always took place at informal dinners where discussions were brief, due to problems of 'time' that made bringing the doctors and the neurologists together difficult. Likewise, the tests for the detection of the cognitive disorders were evaluated and selected on the basis of their simplicity, which meant speediness in administering the questionnaire. And in any case, an effort was made to elaborate techniques to avoid unnecessary digressions during the medical encounter; the invention of the BICQ itself was justified on the basis of the need for a 'simple and speedy' detection tool. The clinical neurologists were critical of the general practitioners' reluctance in adopting their tool. In their view this non-compliance produced more 'superfluous' visits to the neurological clinic, which in turn lengthened waiting lists, increased work loads and thus reduced visiting times. The neurologists attributed delays in the diagnoses of Alzheimer's disease to the general practitioners, not because the latter were considered incapable of recognising the disease but, paradoxically, because they had the 'pretense' to contribute anti-objectifying and personalising observations. The neurologists viewed these negatively because they 'prolonged the duration' of the diagnostic process. The constant search/fabrication for an 'objective' methodology in the diagnostic detection was aimed at saving time, but it appeared to clash with the interpersonal modalities and with the general practitioners' bodily expertise, central to their claim of competence grounded in experience.

'Lack of time' was a common concern to all the actors involved – clinicians, general practitioners, and patients. The general practitioners (whose waiting rooms were always crowded) justified not administering the screening questionnaire due to lack of time. They complained about the bureaucratization of general medicine, which tended to discourage knowledge acquired over the *longue durée* of their historical relationships with older patients. Their conversations with the ethnographer underlined feelings of consternation about the questionnaires. They told the ethnographer that, indeed, they were embarrassed about having to pose standardised questions to people who, whatever the pathological risks may have been, lead 'normal' daily lives – they took their grandchildren to school, did the shopping, rode their bikes and visited their doctors. This embarrassment tended to become even stronger when patients were presented with the BICQ, with some patients responding with rigidity or what the doctors themselves defined as 'anxiety'. The 'anxiety-inducing' effect of administering the test, which clearly is counter-efficient, emerged solely in dialogue with the ethnographer.

From this ethnographic description, a picture emerges of how the internal differences and power relations that underlie the biomedical, administrative and scientific fields correspond to articulations and divisions of the material work needed to construct an object. In this case, the BICQ is made relevant by institutional policy-making choices, and by practices and knowledge applied in the multiple contexts in which various human (patients, family members, general practitioners, clinical neurologists, bureaucrats, anthropologists, and various assistance operators) and non-human (objects, technologies and devices) agencies are activated. This is not to say that Alzheimer's should be deconstructed as an 'invention of the biomedical tradition' founded on 'authoritative' and 'reductionist' ideologies. Clearly, Alzheimer's disease is configured as both protean and concrete, ideal and material, evanescent and, at the same time, resistant, precisely because it has been 'crafted and enacted' (Moser 2008) on a daily basis as second nature within the context of a hegemonic dialectic, whose long-term results cannot be taken for granted.

Concluding remarks

In this paper, proposals for a Gramscian medical anthropology have been furnished, by exploring the pragmatic and analytical efficacy of Gramsci's concept of 'second nature'. The author has tried to demonstrate how the Gramscian political-methodological perspective – attentive to the bodily life of state powers – can be useful in understanding political-physical processes actively involved in the institutional contexts in which early diagnoses of Alzheimer's disease are produced. What is at stake in the dialectic construction and de-construction of second nature is the acquisition of a critical awareness of one's positioning in the process of transformation – which may motivate choices in either acting for change or being against it. The dialectic of constructing second nature embodies thus a series of active relationships, within a network that links people, institutions, technologies and devices. These relationships are not mechanical but organic, multiple and hetero-geneous, inclusive of the human qualities in professional relationships, thus rendering any distinction between ethnography and life experience artificial, and connecting self-awareness to social change. The work of ethnography concerning the fabrication of Alzheimer's disease is responsibly founded on the conscious choice of elaborating critical knowledge of the complex network of contexts and relationships of which the ethnographer himself is one of the crucial points. Through a theoretical practice with a high degree of reflexive and relational agency, the ethnographer's presence may bring to light experiences and knowledge that are practical and bodily, which are overshadowed and not easily communicable in the biomedical field where the existent power relations impose a rigid selection of what is deemed superfluous, unspeakable and impracticable. It is the valorisation of these overshadowed aspects that allows Gramscian 'molecular anthropology' to influence power relations in terms of a social change that is not doomed to fail in any case. In this way an ethnography, which is still in progress, may contribute towards a self-critical understanding of second nature.

Acknowledgements

The author thanks the editors and two anonymous reviewers for their valuable comments, Elisa Pasquarelli and Andrea F. Ravenda for their collaboration in the ethnographic research, which was funded by Regione Umbria: Direzione Regionale Sanità e Servizi Sociali, and Paul Dominici for his help with the English translation. The author followed the ethical guidelines for ethnography of the European Association of Social Anthropologists. Finally, the author acknowledges his debt of gratitude to Tullio Seppilli, and in particular his constant readiness for dialogue.

The paper was presented at the conference 'Medical Anthropology in Europe' funded by the Wellcome Trust and Royal Anthropological Institute.

Conflict of interest: none

Notes

1. For comparison between Gramsci and Foucault in the Italian debate on the philosophy of language, see Lo Piparo (1979, 122n); on the notion of the person, see Gerratana (1997 [1990]); regarding the international debate, see Said (2002) and Ives (2004, 141–4). For an interesting recent comparison between Gramsci's 'common sense' and Bourdieu's 'habitus', see Crehan (2011). For a critique on the rigidity of 'habitus', see Farnell (1999). On the possibility of combining or not combining these critical perspectives for an analysis of relationships between bodies and power, see also Pizza (2004, 2005), Pandolfi (2007), Palumbo (2011 [2008]), and Minelli (2011).
2. On Gramsci's 'living theory', see Pizza (2004 [2003]). In a re-definition of the Italian philosophical tradition known as the 'Italian theory' (and characterised as 'living thought'), the philosopher Roberto Esposito (2010, 178–91) delves into Gramsci's contribution on the question of what we today call 'biopolitics'.
3. Political philosophy has reflected on the notion of nature in Gramsci and on its importance for understanding 'hegemony'. Among the more interesting contributions are Benedetto Fontana's (1996, 2002).
4. The secondary literature on hegemony is extremely rich and multi-disciplinary. Within anthropology, Deias, Boninelli, and Testa (2008) outline a recent debate in Italy; see also the English-language work of Comaroff and Comaroff (1991) and Crehan (2002). For a genealogy of sources on this question, see Boothman (2008), Fontana (1993, 2002).
5. The author's translation. SPN: 286 translates it as 'a new type of man'.
6. Giovanni Pizza (co-ordinator), Elisa Pasquarelli and Andrea F. Ravenda make up the team of anthropologists.
7. Over the last 30 years the anthropology of Alzheimer's has taken on various approaches (Leibing and Cohen 2006). Initially it centred on the concept of the 'person' with the phenomenological-cultural approach. More recently, another interpretation of Alzheimer's is beginning to break ground, which places greater attention on the policy-making and power relations inscribed in the scientific and laboratorial procedures (Moser 2008; Behuniak 2010). The latter is closer to the perspective proposed in the present study.
8. This is a rather controversial recent diagnostic. See among others Moreira and Palladino (2008), Whitehouse and Moody (2009).

References

Behuniak, S.M. 2010. Toward a political model of dementia: Power as compassionate care. *Journal of Aging Studies* 24: 231–40.

Boothman, D. 2008. Hegemony: Political and linguistic sources for Gramsci's concept of hegemony. In *Hegemony. Studies in consensus and coercion*, ed. R. Howson and K. Smith, 33–50. New York: Routledge.

Cirese, A.M. 1976. *Intellettuali, folklore, istinto di classe. Note su Verga, Scotellaro, Gramsci.* Torino: Einaudi.

Comaroff, J., and J. Comaroff. 1991. *Of revelation and revolution: Christianity, colonialism and consciousness in South Africa*, Vol. I. Chicago: University of Chicago Press.

Crehan, K. 2002. *Gramsci, culture and anthropology.* London: Pluto Press.

Crehan, K. 2011. Gramsci's concept of common sense: A useful concept for anthropologists? *Journal of Modern Italian Studies*, issue *Gramsci revisited. Essays in memory of John M. Cammett* 16, no. 2: 273–87.

Deias, A., G.M. Boninelli, and E. Testa, eds. 2008. *Gramsci ritrovato. Lares. Rivista quadrimestrale di studi demoetnoantropologici diretta da Pietro Clemente* LXXIV, no. 2. Firenze: Leo S. Olscki.

De Martino, E. 1949. Intorno a una storia del mondo popolare subalterno. *Società* V, no. 3: 411–35.

De Martino, E. [1961] 2005. *The land of remorse. A study of southern Italian tarantism*, translated and annotated from Italian by D.L. Zinn. Foreword by V. Crapanzano. London: Free Association Books.

Esposito, R. 2010. *Pensiero vivente. Origine e attualità della filosofia italiana.* Torino: Einaudi.

Farnell, B. 1999. Moving bodies, acting selves. *Annual Review of Anthropology* 28: 341–73.

Fiori, G. 1990. *Antonio Gramsci: Life of a revolutionary.* London: Verso.

Fontana, B. 1993. *Hegemony & power. On the relation between Gramsci and Machiavelli.* Minneapolis: University of Minnesota Press.

Fontana, B. 1996. The concept of nature in Gramsci. *The Philosophical Forum* XXVII, no. 3: 220–43.

Fontana, B. 2002. Gramsci on politics and state. *Journal of Classical Sociology* 2, no. 2: 157–78.

Frankenberg, R. 1988. Gramsci, culture, and medical anthropology: Kundry and Parsifal? or rat's tail to sea serpent? *Medical Anthropology Quarterly* 2, no. 4: 324–37.

Gerratana, V. 1997 [1990]. *Gramsci. Problemi di metodo*, Roma: Editori Riuniti.

Giaquinto, S., and L. Parnetti. 2006. Early detection of dementia in clinical practice. *Mechanisms of Ageing and Development* 127: 123–8.

Gramsci, A. 1975 [1948–1951]. *Quaderni del carcere*, critical edition by V. Gerratana, Torino: Einaudi. English translations: *Prison notebooks*, ed. J.A. Buttigieg. New York: Columbia University Press, Vol. 1 (1992), Vol. 2 (1996), Vol. 3 (2007); *Selection from the prison notebooks* [SPN], ed. Q. Hoare and G. Nowell Smith, London: Lawrence & Wishart (1971); *Further selections from the prison notebooks* [FSPN], ed. D. Boothman, London: Lawrence & Wishart (1995). (Original written between 1929 and 1935.)

Gramsci, A. 1994 [1947]. *Letters from prison*, ed. F. Rosengarten, transl. R. Rosenthal, New York: Columbia University Press. Original edition: *Lettere dal carcere*, Torino: Einaudi, 1947; new edition: *Lettere dal carcere*, ed. S. Caprioglio and E. Fubini, *Introduzione* and *Note* by S. Vassalli, Torino: Einaudi, 1965; further edition: *Lettere dal carcere 1926–1937*, ed. A.A. Santucci, 2 volumes, Palermo: Sellerio, 1996. (Original written between 1926 and 1937.)

Ives, P. 2004. *Language and hegemony in Gramsci.* London: Pluto Press.

Leibing, A. and L. Cohen, eds. 2006. *Thinking about dementia. Culture, loss, and the anthropology of senility.* New Brunswick, New Jersey: Rutgers University Press.

Lo Piparo, F. 1979. *Lingua, intellettuali, egemonia in Gramsci.* Bari: Laterza.

Minelli, M. 2011. *Santi, demoni, giocatori. Una etnografia delle pratiche di salute mentale.* Lecce: Argo.

Mol, A. 2002. *The body multiple. Ontology in medical practice.* Durham, NC: Duke University Press.

Moreira, T., and P. Palladino. 2008. Squaring the curve: The anatomo-politics of ageing, life and death. *Body & Society* 14, no. 3: 21–47.

Moser, I. 2008. Making Alzheimer's disease matter. Enacting, interfering and doing politics of nature. *Geoforum* 39: 98–110.

Palumbo, B. 2011 [2008]. La somiglianza è un'istituzione. Classificare, agire, disciplinare. In *Saperi antropologici, media e società civile nell'Italia contemporanea*, ed. L. Faldini and E. Pili, 207–45. Atti del 1° Convegno Nazionale dell'ANUAC, Matera, 29–31 maggio 2008. Roma: CISU.

Pandolfi, M. 2007. Il corpo: Resistenza soggettiva, resistenza collettiva. In *Il tessuto del mondo. Immagini e rappresentazioni del corpo*, ed. F. Faeta, L. Faranda, M. Geraci,

L. Mazzacane, M. Niola, A. Ricci and V. Teti, 122–41. Napoli-Roma: L'Ancora del Mediterraneo.

Pizza, G. 2004 [2003]. Antonio Gramsci and medical anthropology now. Hegemony, agency and transforming persons. *AM. Rivista della Società italiana di antropologia medica* 17–18: 191–204.

Pizza, G. 2005. *Antropologia medica. Saperi, pratiche e politiche del corpo*. Roma: Carocci.

Pizza, G., and H. Johannessen, eds. 2009. Embodiment and the state. Health, biopolitics and the intimate life of state powers. *AM. Rivista della Società italiana di antropologia medica* 27–28. Lecce: Argo.

Said, E. 2002. In conversation with Neelandri Battacharya, Suvir Kaul, and Ania Loomba. In *Relocating postcolonialism*, ed. D.T. Goldberg and A. Quayson, 1–14. Oxford: Blackwell.

Samuelsen, H., and V. Steffen. 2004. The relevance of Foucault and Bourdieu for medical anthropology: Exploring new sites. *Anthropology & Medicine* 11, no. 1: 3–10.

Santucci, A.A. 2010. *Antonio Gramsci*. Preface by E.J. Hobsbawm. Foreword by J.A. Buttigieg. New York: Monthly Review Press.

Whitehouse, P.J., and H.R. Moody. 2009. Mild cognitive impairment. A 'hardening of the categories'? *Dementia* 5, no. 1: 11–25.

Of divinatory *connaissance* in South-Saharan Africa: the bodiliness of perception, inter-subjectivity and inter-world resonance

René Devisch

Institute for Anthropological Research in Africa, KU Leuven – University of Leuven, BE-3000 Leuven, Belgium

Branching out from recent perspectives on divination in Africa, this study explores a fresh approach that engages in a constructive dialogue between local knowledge practices and Western-derived human sciences. A first section positions this essay within an emerging debate over the perspectival ontological turn in anthropology – in line with Viveiros de Castro (2004) – which holds that people's culture-specific ontology – such as, envisaging some propensity in the fabric between the human, things, fauna, flora and inter-worldly force-fields – is most explicitly voiced in the divinatory oracle and expressed in the ensuing healing and societal redress. The study then outlines the post-Lacanian matrixial model, defined by Ettinger (2006), that re-examines the originary processes unlocking, and inter-connecting in, the matrixial borderzone between body and psyche of mother and foetus or *infans*. A second section then focuses on the oracular scrutiny typically employed by the mediumistic Yaka diviner in southwestern DR Congo. Such practice, it is contended, induces the diviner to sense out in the consulting kin group the bewitching force-effects and the unspeakable in the inter-generational realm. The oracle unconceals some unguessed fate in the client's inter-generational line, in particular the inter-corporeal embeddedness of latent memory traces and forces of ill-bearing. Third, the study will conclude by evaluating – along the terms of the local culture's genius – the perspectivist stance and matrixial model.

The worldwide dissemination of geomantic and shamanic divination from Palaeolithic times across and beyond Asia and Africa counts as one of the oldest forms of proto-globalisation. Indeed, this millennia-old institution comprises techniques of increasing formalisation regarding society and the propensity of things in the universe. There is a growing body of historical and comparative research on divination, and some of the main perspectives on its global cultural history have been explored or co-ordinated by Wim van Binsbergen (2003, 2011). Divination today surfaces in different local forms in many parts of the world, as the vast study by Langer and Lutz (1999) so well shows.

As a complement to an earlier survey (Devisch 1985) on the major anthropological approaches to divination in Black Africa, a follow-up review essay (Devisch, 2012) assesses recent experiential-cognitive approaches to divination, understood as a local form of oneiric, highly intuitive and acute bodily-sensory perceptiveness and knowledge production. The review moreover sketches current phenomenological views that approach divination as a means to achieve unprecedented insight and management of relationality and, in urban contexts, of inter-cultural knowledge-sharing.

Branching out from recent perspectives on divination in Africa, this study explores a fresh approach that aims at a constructive dialogue between Western-derived human sciences and local knowledge practices. It seeks to address and refute some modernist ethnocentric prejudices or reductionist views on divination. There is no intention here of reducing geomantic or shamanic divination's keen perceptiveness to some mode of so-called mystical or paranormal scrutiny. Divination, understood in its own terms, is a process involving complex – bodily, affect-laden, sensory, intuitive, mediumistic, artistic – folds and skills of emotional perceptiveness and feel-thinking. Divination, it is argued, delves into sealed layers of empathic consciousness of the client and consultants, while sensing out the unthought-in-thought, the unspeakable or undisclosable in the client's history and that of the family. Divination is fundamentally a quest into the unrepresentable or unspeakable, which lies buried in one's shadowy side or unconscious otherness of self. Similarly, it embraces some unguessed fate along the transgenerational line of inter-corporeal, inter-subjective and inter-world connectivity, qualifying the consultants' family.

A note on fieldwork (1972–2003)

Before turning in the next section to the contextual practice of divination in the *yiYaka* speaking society in the Kwaango region of southwestern DR Congo and Kinshasa's shanty towns,[1] let us briefly define the applied perspectivist and matrixial approaches that help to pursue fresh insight into the intertwining of body-self, desire and affect in the divinatory consultation and in people's other health-seeking practices. Not only the data but also the author's experiential perspective derive from his fieldwork carried out from 1972 to 1974 among the Yaka of northern Kwaango along the Angolan border, in the Taanda settlement of 13 villages – some 120 inhabitants in average, or six persons per square kilometre – located 450 km to the south of Kinshasa (Devisch 1993, 2011; Devisch and Brodeur 1999). The investigation gained further momentum through collaborative work with Congolese and Belgian colleagues for annual medical-anthropological research ventures (1986–2003), which spanned almost three weeks or more among the Yaka and Koongo people residing in a number of Kinshasa's shanty towns. During *in situ* supervision of doctoral research in a variety of African countries since the 1990s, the author has been able to witness a number of divination séances among the Kasena, Igbo, Gusii, Zulu and Xhosa.

Perspectivist approach and matrixial model

People's focus on the co-resonance between worlds has been central to the anthropological efforts that the author has undertaken for four decades regarding Yaka people's socioculture. This predominantly medical anthropological research spurred increasing reflexivity on his empathic research involvement in local people's anxieties about health, and in their health-seeking activities informed by their divinatory consultation. The concern has been how to do this rethinking in accordance with the very perspective and cultural genius of the subjects. Additionally, in seeking to overcome the imposition on Yaka practices of some Western modernity's polarising thinking – along such dualities as mind/body, reason/sense, rational/irrational, modernity/tradition, developed/underdeveloped, culture/nature, literacy/oralcy – the author has come to favour the experiential and inter-subjective phenomenology of Merleau-Ponty (Devisch and Brodeur 1999), and the related perspectivist stance along the lines of Viveiros de Castro (2004), as well as the matrixial feel-thinking model elaborated by Ettinger (2006). The aim is now to connect with the diviner's and consultants' shared and vivid experience of being-in-the-world, which is qualified by the ill-bearing, or the health-enhancing, inter-corporeal, inter-subjective and inter-world reverberation and inter-connectivity effected by the oracle (and subsequent healing and social redress).

For the author, the just-mentioned inter-subjective and inter-corporeal mind-fulness alerting to affliction and trauma resonates with his family's durable inscription in the particularly traumatic experience that their generation of Flemish-speaking people had in Belgium during both World Wars. In his attentiveness to afflicted people and the aetiology of affliction, as well as healing and appeasement, both the Yaka and Flemish people strongly resonate *through him*. Acknowledging well that Yakaland and Flanders are worlds apart, he however has not stopped acting as the matrixial borderlinker who resonates with sensitivity to the matrixial in each through their extimacies, in particular his own. Extimacy – a Lacanian notion – represents a shadowy side or unconscious otherness at the very core or intimacy of self, in both the host and the researcher.

The *perspectivist* approach draws on the treatment of the body-subject in Merleau-Ponty's (1964) later writing, and on his analysis of plural, acute and elucidating perception that starts from the pre-reflective experience of being-in-the-world. In perspectivism, we come to perceive our life-world from the perspectives of our fellow-subjects and those in tune with the significant inter-animating agencies. It appears that Merleau-Ponty's approach helps to better understand the diviner's situated grasp of the client as body-subject or body-self. It further allows for insight into the diviner's sense of the client's world at the level of 'flesh' (*la chair*) as a tissue of sensible and desirous being-in-the-world. In this view, perception is in fact the perceiving subject being herself a part of the world.

A perspectivist anthropological attention for the culture-specific views on the co-substantiality and co-resonance of life-forms helps to punctuate in a variegated way the Yaka ontology that envisages some propensity in the weave between the human, things, fauna and flora, as well as the inter-worldly force-fields. The latter, being more conventionally labelled as interdependence with the otherworldly, refers to the given people's attunement to their deceased members, ancestors and spirits, deities or divinities, arcane or bewitching powers, as well as Satan proper to the Christian imaginary. Contentions of the culture-specific ontology are most explicitly submitted to and voiced in the divinatory oracle, or taken care of in the healing, as well as in funerary and other life-cyclical rituals. In these, the reproductive, communicational and transformative acts or movements in the group and life-world – sexual, life-cyclical, alimentary, musical, verbal, ritual, agricultural, seasonal, lunar, solar – are portrayed, hence playing out some fundamental in-betweenness or tension qualifying the unceasing fabric of both the life of people and of their individual, collective and inter-world fates. And in line with this ontology of co-substantiality among all life-forms, whatever a hunter, priest, elder, cook, craftsman, or other actor approaches from his or her point of view may demarcate a kind of anthropomorphic interplay between levels and states of reality. The fabric nests the unfathomable and the unconscious or shadowy otherness of self at a core of experiential reality that, in particular by way of ritual, happens to be activated, 'agented' or personified into 'someone' by the point of view of the healer-priest, diviner, initiate, elder. This personification evokes a *subjectification of objects*, in line with what Edouardo Viveiros de Castro (2004) brings out regarding the Amazonian anthropomorphic animism or perspectivist multinaturalism. For example, Yaka healers and initiates tend to grant a form of subject position not only to the spirits but also to the ritual power objects they make – ethnocentrically labelled as fetishes – while considering these as inhabited by a spirit, if not by the maker's or user's well-wishing, or malevolent, intention.

Individual development, kin relations and societal organisation, as well as transformative intervention in the realm of 'nature', are inseparable from the *local perspectivism* on a particular record of sensed out or lived in phenomenal, in particular sociocultural, reality. The experiential realm of 'nature' is, for people in Kwaango land, of a great plurality given the singular specificity of affects, capacities and dispositions of the bodiliness or of each subject and of each animal species and interacting substances involved. Yaka socioculture sees the human subject very much as both a desiring author and a steady scene of exchange, namely of inter-corporeal, inter-subjective and inter-world transactions between their respective fields in resonance (Devisch 1993: 53ff.). The anchor point here for the subject is corpo-, socio- and cosmo-centrically formed in the differing contextual practices of gift-giving, relationality and inter-animation or mutual enlivening. On the level of a subject's body, the sensual skin with its sensory capacities, the bodily 'cavities' with vital organs (in particular the hollow bones, the heart and liver in the 'body's furnace', the genitor's semen stocked in the bony skull and spinal cord, and the womb and stomach in the belly or 'body's kitchen') form a most vital borderzone. Next to these sensory capacities and bodily cavities, the articulations of the limbs constitute the heavily laden junctures on which most vital connections and exchanges with others and with the visible and invisible worlds are being projected and handled. Moreover, sensory capacities, orifices, cavities and articulations of the body are crucial junctures for most significant 'borderlinking' interventions – both unblocking and reconnecting via massage, vegetal ointments, tonifiers, infusions, enemas – in the subject's physical body, the body-self, the group and life-world. The culture's attention goes to the transactions not only at the body orifices and skin level, but also to those at the doors and borders of the home and homestead, as well as to the crosspoints and transitional zones and critical moments in the life-world.

Communal life in the local homestead smoothly inserts itself as an integrative force in the larger environment with all its life-forms. When expressed in the perspectivism of Yaka people, its matricentric concern or point of view is given in the specificity of human bodies. To say it in line with Viveiros de Castro (2004, 475–6), each body with its 'innate nature' appears as a singular assemblage of affects, capacities, agency, and ways of being and interconnection, that are the very base of perspectivsm. At the same time, the body as a fact of the 'innate nature' posits its inherent inter-species transformability, or may inspire the self-bewitching fear of seeing the animal in the human, or the – bewitching or deceased – human within the body of the animal that one encounters, dreams of, or eats. This multinaturalism – Viveiros de Castro's (2004) expression – is the Yaka opposite to the modernist Enlightenment nature/culture, animal/spiritual, reason/imagination or visible/ invisible divide. Congruently, in the Yaka perspective, the 'village space' (*hata*) literally means a space 'cut out' (*-hata*) of, and being part of, the more encompassing 'forest realm' (*n-situ*). In other words, for the Yaka imaginary, society is an integral part of the larger environment, or, translated in modernist terms, culture is a part of nature. In people's view, the balance between an emotional pull in the individual's heart, or her 'intuitive feel-thinking or *connaissance*' (*-zaaya*), as distinct from the more 'reflective and thus responsive thinking' (*-baandza, -yiindula*) that draws on 'mirroring insight in the heart' (*-mona mu mbuundu*), are qualities ascribed respectively to the female or male elder. The Yaka do associate the maternal with primordial 'nature' – not the modernist notion of instinct or carnality, but in the sense of the paradigm of the cyclical life-spring and processes of transformation, transfer or exchange. And 'the exchange model of action supposes that the subject's "other" is another subject', i.e. a mode of subjectification, hence not an object (Viveiros de Castro 2004, 477).

For sure, Ettinger's *matrixial* psychoanalytic theory (Ettinger 2006), in as much as it concerns forms and modes of responsive and life-bearing relatedness, shared borderspace and transferential borderlinking, may inspire a heuristic perspective into inter-species transformability. At play here are *borderlinking* practices that unveil, unlock or unsettle in the subject or the life-world unsuspected sources of empowerment, modes of meaning and interpretational possibilities, while they simultaneously interconnect or intersect with, and hence remould, the new perspective or figuration, inter-locution or comprehension, way of seeing, value or mode of empowering. In the view of Yaka people's ontology, spirits and the deceased, priests, or healers and other initiates – depending on their proper initiatory conditioning of intention or perspective – can embody or interconnect with some properties of totemic animals, plants or objects. Ettinger's theorising regarding the '*matrix*', or the pre- and post-natal *infans*-mother co-resonance as it is most readily grasped by way of feel-thinking and co-implication, has proved very useful in the effort to properly understand Yaka people's deep sense of inter-world attunement and empathic osmosis with the invisible. Her theory offers insight into the imperceptible transfusion of states of being, or into the sensual inter-corporeal and inter-subjective openness to one another's fantasies and desire, affects, longings and emotions, sense of both the good and beauty. It reports modes of transfusion or openness of beings largely beyond the grasp of the optical and conceptualisation.

Among the most noteworthy of the healing cults' matrixial focus on the body-self is their feel-thinking in line with passageways, encounter, sharing and folds; and joint transmission in the vital (inter-)corporeal functions. These include the heartbeat and circulation of blood, menstruation cycle and reproductive sexual union, pregnancy and parturition, the sharing of the meal and digestion, and – above all – breathing and its audibly accelerated or decelerated rhythm. They also include women's dancing at social or seasonal transitional moments. The Yaka associate the advent of life or of death with some irresolute beginning or cessation of audible breathing. In their view, the invisible primordial chthonic womb – in its numinous or augury dimension made present in the home of an initiate or a lineage patriarch by a clump of kaolin-like clay – is the local world's and society's innermost cradle that fundamentally energises the indirectly perceptible visible and self-steering vital bodily processes of the dependants, giving them rhythm and form. Indeed, the matrixial processes introduce in each body-self a differential bodily economy, such as in the heaving of the chest in breathing, the heartbeat, the digestive process, the menstruation cycle and auspicious foetal development, as well as keeping these processes and their cycles in resonance with the social and cosmological bodies. What is invisible and intangible – in the bodily, social and cosmological economies – can assist, particularly in the work of cults, to disclose the unspeakable in the visible.

Among the Yaka, divination and its clients see the otherworldly realm as underpinning the 'thisworldly' physical and social-symbolic order, and as holding the key to the individual's ability and good or bad fate (Devisch and Brodeur 1999, 93–117). Yaka cosmology, in its matrixial scope, underscores people's sense of the world and human life as ceaselessly emerging but fundamentally in-appropriable. The world and human life are depicted in their never-ceasing emersion of capabilities, however versatile, like the never-failing transition between day and night, sun and tempests, life and death, satiety and hunger, or goodness and evil in each of us. In the case of a lasting disempowerment or affliction (insoluble conflict, disaster, exceptional loss, lack of income, chronic illness, misfortune, threat of death, or difficult decisions concerning the foundation of a home, marriage or divorce, or migration opportunities), Yaka people in uncertainty may call upon the mediumistic capacity of a diviner to lay bare the cosmological and social issues bearing on the problem.

Nature and role of the Yaka mediumistic diviner

This paper deals with diviners who participate in the same inter-regional Ngombu cult of affliction, indeed a major legacy of Bantu civilisation. Yaka procedures of divinatory consultation and oracular scrutiny – like those of the initiation of a diviner-to-be, who may be male or female and acts androgynously – have been carefully documented in the early 1940s by the Jesuit missionary De Beir (1975, 114ff.) and later recorded audiovisually by Dumon and Devisch (1991). A former study (Devisch and Brodeur 1999, 93–123; Devisch, manuscript) reported how the Yaka organise and comprehend the initiatory moulding and consecration of the diviner's acute perceptivity, as enacted in the Ngombu cult. The intention here is to provide a perspectival and matrixial analysis that closely follows the rationale of the local culture's logic and ontology, as well as the group's basic concerns.

A divinatory consultation

In the event of an accident, illness or other affliction, the consultants – representing the various duties and rights in the afflicted kin group – seek out a diviner who, according to convention, lives at least one day's walk from the home of the client; this step ensures that the consulted diviner knows nothing by hearsay of the concrete circumstances of the affliction and is unlikely personally to know the client or kin group to any extent. The term client refers to the individual who has recently become deceased, or is actually ill or afflicted, but who is only present at the oracle in the case of a travel project or a problem with the hunt; the corpse is never physically brought to the diviner.

Shortly after the arrival of the consultants, the diviner lithely dances to the rhythm of the slit-gong she plays, thus marking the beginning of the preliminary phase of the consultation. She then places her divination basket between her and the consultants and from time to time sniffs the client's double (an object placed the night before above the client's heart) or holds it in front of a small mirror. Her bonds as a diviner with 'the primordial chthonic womb of life' (*ngoongu*) are enlivened by white clay smeared on her left wrist. Crouching over her gong – its slit turned towards the consultants – the diviner adopts a position similar to that of a woman in childbirth.

The first phase of the consultation proper is indicated by the diviner's beating of her gong and the interrogation of the oracular grid. For this purpose, she employs archaic mythical language, which she chants in a monotone voice, called -*kedibila*, literally, interpolate or debate. This language sets the scene for the 'unconcealment' of the unspeakable in the client's consciousness and life-world. The barely intelligible mythical terms are taken up, by way of refrain and in an enchanting tempo, by the consultants in chorus. While the consultants are drawn into the process and even incited to do some introspection, the esoteric wording of the chant provides no factually significant or concrete information at this stage. Totally unassisted by the consultants, the diviner must grasp at least the gist of the concern the consultants have brought to her, relying solely on her dreamwork and acute flair. Evoking some concrete yet unspecified signifiers (such as 'I see a bereavement'; 'I see a man wounded at the hunt'; 'Cursing has tied shut your wife's uterine life flow'; or 'Bewitchment has been fatal at your home'), the diviner begins to divulge her initial reading of the case. Tentative statements or markers of problematic elements in the case such as these, set against a mythical background, at this point receive no further contextual clarification or application.

If satisfied with the proceedings so far, the consultants may request that 'the oracle further sorts out what is on the body of the deceased'. With their tentative agreement and payment at this stage, along with their growing expectations, the consultants enact the co-implication developing between them and the diviner. This interaction furthers the transference relation before initiating the aetiological oracle proper, which usually occurs the following morning. Then 'to catch or seize the problem' the diviner has to rely on her 'acute flair or sharp nose' and 'seeing in dream' inspired by the deceased diviner-predecessor.

The consultation gropingly unfolds – customarily, the next morning – into a scant aetiological scrutiny of the case along the basics of the local world order. As the oracle (reported in Devisch, 2012, 91–3) shows, it gradually inscribes the consultants' concern into the self-legitimating space-time of authoritative divinatory unconcealment.

The tissue of the diviner's desirous, sensible and heightened perceptiveness

Aetiological scrutiny

Diviners insist it is the slit-gong itself that utters the oracle. The gong acts as the diviner's 'double' (*yilesi*) and is 'her very effigy'; this forces the diviner to turn onto herself and to become a palpating touch, eye and ear – both a tangible, visible and audible entity and at the same time a touching, seeing and speaking agency. It is the phallic stick, placed slantwise in the slit of the gong and used to drive the rhythm of the oracle by beating the lips of the slit, that creates the link between the pre-eminence accorded to the mother and the uterine, the pulsional, com-passionate and transferential on the one hand, and difference and differentiation, articulated in particular by the ancestral, agnatic and paternal traditions on the other. The oracle proper is now seen, from the aetiological point of view, to displace the diviner from centre stage: she is no longer considered to be the subjective source or self-reflective author of the enunciation. This disclosing function is now associated with the Ngombu spirit and the primordial uterine life-source, *ngoongu*, speaking through the womb-like slit-gong, all the while situating the consultants' concern within the life-bearing versus life-harming potential of the family.

The dynamic of embodied identification with the slit-gong brings the diviner's gong close to operating what could amount to some radical perspectivist ontological move; it evokes the process of *'Thinking through things'* (Henare, Holbraad, and Wastell 2007), such as performed by the African-American shaman in Cuba, among others, whom Martin Holbraad has described. The oracle has now become the voice of the self-reliant uterine life-flow from which the client and those concerned stem, and into which they tap, in both good and ill health. It thereby reveals the dual affect-laden logic undergirding the community's axiological or moral order of sociality: namely the ethical law of exchange, as it bears on the family, in the matriline, and thus on life's socially-constructed fabric, whether conducing these to flourish or be sucked out in bewitchment (Devisch and Brodeur 1999: 93–154).

Divinatory aetiology is grounded in the principle that anything inhibiting or obstructing life may ultimately be attributed to one or another instance of transgression against the order of basal reciprocity. Diviners reckon this as a form of 'theft' (*-yiba*), the quintessential notion of evil, representing the most fundamental denial of this order of reciprocity. It is a life-disabling or life-threatening act, or an ill-intentioned disregard or infringement of the law of exchange in the family group – and more specifically in the uterine line of filiation. An oracle proceeds first to identify some basic transgression and the retaliative bewitchment or curse it elicits, as well as the ancestral wrath it in turn may have provoked. In doing so, the oracle transforms a crucial violation of the client's vital weave into a localisable, specific and manageable entity. The oracle is then followed by a family council, whose task it is to negotiate any steps that need to be taken in order to avert evil and bring healing to the afflicted person/s, and ultimately to achieve the renewal of the social and cosmic fabrics. In this context the maternal uncle is a key figure, and his status is such that he can both prevent or authorise life-taking within the uterine group. In stark contrast to other discourses of healing, throughout this process none of the actors – whether diviner, oracle, family council or maternal uncle – are perceived to be salvific agencies in pursuit of some higher Truth or of The Good. It is only within the intricate complex of affects and life-bearing or life-taking forces that the maternal uncle, like the family council, may in fact be regarded as a highly ambivalent agent capable of shifting roles, from that of a generous and vital corroborating co-implication to that of an abusively self-serving or maleficent actor.

Disclosure of the client's fate

For the Yaka diviner, much as for the divinatory practice witnessed in other parts of Africa, the focus of the oracle – and the object of the diviner's intention and scrutiny – is on life's basic concern with the inescapable and sustainable reciprocity dictated by the culture's moral order. The root cultural model for such reciprocity is that of the mother-child dyad and the uterine blood bond. In the Yaka context, the moral order bears the self-evident status of *k'amba,* which literally means 'is it not so that', 'is it not evident that'. This order alludes to the propensity of the things of life-transmission across generations and to the doxa of inescapable or assigned fate, next to the unfathomable blanks in the surreptitious trans-generational transmission of trauma, preferences or avoidances and secrets.

Almost every element, object or movement in Ngombu divination – the slit-gong, the kaolin or the diviner's dance – fundamentally relates in Yaka perspective to modes of emerging, breeding or self-renewing forms of being – and hence, of shadowy, untamed or passionate forces as well. As argued in a detailed analysis (Devisch, 2012, 90–5), these forms or forces are implicitly evoked in the popular imagination as well as in the diviner's initiatory seclusion by the brooding hen, orgasm, gestation, parturition or spirits, the night, death-agony and trance-possession. They haunt men's and women's conceptions, and motivate their attraction for what reaches beyond the order and limits of their known and domesticated world: including invisible and untamed forces, unthought sources of desire, worry and sadness, as well as sleep or dream, arousal or anger, enthusiasm or anxiety, cursing or witchcraft. In other words, they ply the collective imaginary with unanswerable questions or dark holes and blanks, or the unthought-in-thought.

The notion of 'shadow' is key to understanding extimacy and the Ngombu divinatory perspective. The disturbance effected by any misfortune or family trauma occurs, or rather unleashes itself, in the 'shadow' (*yiniinga,* literally, something that swings at, *-niinga*) of the afflicted, and it somehow destabilises her body borders, inter-corporeity and inter-subjectivity. A complex entity, the shadow or double always accompanies yet is extimate to the subject; very much like her dreaming, the shadow projects the subject beyond her bodily space-time limits. In the Yaka view, the extimate – intimate but alien – identity element portrays the given person's unspeakable wishes, irascible refusals or denials, unruly drives or impulses, and unfathomable desires. Now it is precisely these drives and desires in the subject that are so often blocked or disputed by the desires of others; as such they then become identified as envy, greed and intrusion, or spirits and witchcraft – all invisible. In the Yaka perspective, the unrepresentable invisible allows itself to be imagined at the fluid border of what can be perceived yet tends to be hidden, such as in the *yiniinga* or shadow.

A subject's extimacy is highly vulnerable to the effects of the webs and plots of the personal or family imaginary in the past and present, and it is easily haunted by unsettling experiences or undisclosable – yet easily rekindled – inter-generational memory images and traces. Such traumatic memories or experiences likely involve some intuitive discernment of dreams or trance, a startling event, or violent words such as cursing, or even intrusions or transgressions in the presence of the afflicted. Once they have been identified, such images or traces serve to guide the diviner gropingly to explore and enunciate, by way of mere evocative signifiers, the shadowy dimension in the client and others concerned. The shadow offers itself, in the subject's dreams or heightened experience, as something extimate to the self inasmuch as it may help her to express some grasp – beyond reason – of a desire shared with a complicit other, or some awareness of an unconscious trauma that may still trouble family members. At the same time, that shadow may either disclose or conceal the holes in the subject's and family's often tacit awareness of their respective inter-subjective, inter-generational and inter-world weave.

Conclusion

This research has led the author to adopt the *perspectivist* approach, which allows for the full valorisation of people's contextual ontology at play within the processes of divinatory consultation, as in the initiation, hence also in people's outlook on the force-fields they are in.

Rather than viewing divination merely as a mode of extrasensory clairvoyance, Yaka culture clearly regards the diviner as a *medium*. The diviner is able to track down the problem and 'unconceal' the not-yet-thought-out in the case submitted to the oracle by virtue of her acute perceptive sense, keen intuition and divinatory skills. In the Yaka worldview, the voice of the slit-gong – gropingly uttered by the border-linking diviner – is granted an authority above and beyond that of any human being to act as a medium, of both the divinatory oracle and the realm of the unspeakable within the family reality. The diviner's disclosing words, tapping into her dreamwork and gifts of scrutinising flair, and originating from the slit-gong and the primordial uterine and chthonic life-source, would appear to sense and flush out the inter-generational tracks of envy, malignity, retort or bewitchment in the family. The diviner and the clients, or even the researcher, appear as 'flesh' of both body and world; that is, they are drawn, as sensible and desirous beings – as they wish to be touched, seen and embodied – into the things of the world. We then find that the diviner's multi-sensate participation in the unconcealment is one of espousing the pulsional and sensible yet muted or shadowy depth of what is being unconcealed. And it is this insight that provides us with an innovative point of departure for theorising the diviner's acute perceptiveness of the latent inter-generational memory traces and the imaginary plots at play in the client's and family group's relations and life-world.

Divination, as observed within Yaka and other South-Saharan contexts, deals first and foremost with existential realms that are replete with forces and pervade all of human reality. The notion of forces begs for a more intimate and contextually-sensitive mode of comprehension, and this is provided by a genuinely *matrixial* interpretational perspective. The matrixial approach requires a sensuous and rhizomatic, corporeal and sublinguistic opening-up to the oneiric, pulsional and affective streams of inter-corporeal motivations and inter-subjective moods or messages. It implies an intuitive feel-thinking of resonance, and even inter-communication between the human, things, fauna and flora, ancestors, spirits and bewitching forces, whether they are visible and tangible or not.[2] Inasmuch as this matrixial model has been useful in dealing with the inter-generational and inter-subjective 'response-ability' to the traumatic hieroglyphs of one's life-world, it is invaluable in achieving an adequate interpretation of the diviner-medium's trans-world utterances from the uterine slit-gong and the primordial chtonic womb of life.

Acknowledgements

The research was financially supported by the Belgian National Fund for Scientific Research, the Research Fund – Flanders and the European Commission Directorate-General XII. The author has followed the ethical guidelines for ethnography of the European Association of Social Anthropologists. Peter Crossman provided invaluable editorial assistance.

Conflict of interest: none.

The paper was presented at the conference 'Medical Anthropology in Europe' funded by the Wellcome Trust and Royal Anthropological Institute.

Notes

1. The Kwaango region's rural-based society numbers some 350,000 speakers of *yiYaka* in the vast expanse of the southern savannah on the Angolan border. The domestic and public day-to-day life-scene of great scarcity is one of hunter-gathering and small-scale subsistence farming. It is a society historically devoid of any autochthonous statist edifice, monotheist religion or overall master discourse. The Yaka are widely reputed for their divinatory art and healing cults. They combine a social identity derived from patrilineal descent with the notion of uterine filiation of the 'innate nature' originating with the very source of physical life, traced back to the maternal great-grandmother. While over the last decades half of the Yaka population have taken advantage of elementary school education, we should acknowledge that, in DR Congo, rural areas are increasingly bereft of good educational opportunities which only a few privileged city schools offer. It is reasonably likely that an equal number of people today define themselves as Christian, having been baptised at the school-going age or having at a later age joined a neo-pentecostal church. The state logic from the administrative centres creeps in very slowly in the Kwaango region, mainly through the educational and religious persuasions, the back and forth movements of persons, and the mercantile rationale and influx of basic consumer goods.
2. Ngombu initiation, for example, effects a transworld passage, evoking the behaviour of the otter-shrew which goes underground before re-emerging, thereby evoking rebirth in the manner of brooding (see Devisch, 2012, 89–95). The otter-shrew is an amphibious animal that is a notorious chicken-killer; and a chicken may eventually function as a sacrificial equivalent to the human being. The diviner's initiation thus produces an ontological change in both the diviner-novice and her slit-gong, which is her very double.

References

De Beir, L. 1975. *Religion et magie des Bayaka*. St. Augustin-Bonn: Anthropos.

Devisch, R. 1985. Perspectives on divination in contemporary sub-Saharan Africa. In *Theoretical explorations in African religion*, ed. W. van Binsbergen and M. Schoffeleers, 50–83. London: Routledge.

Devisch, R. 1993. *Weaving the threads of life*. Chicago: University of Chicago press.

Devisch, R. 2010. Of divinatory connaissance among the Yaka of Congo. Manuscript.

Devisch, R. 2011. Online lists of publications, available at: https://perswww.kuleuven.be/~u0012668/; http://www.iara.be/

Devisch, R. 2012. Divination in Africa: The bodiliness of acute discernment. In *The Wiley-Blackwell companion to African religions*, ed. E. Bongmba, 79–96. Oxford: Blackwell.

Devisch, R., and C. Brodeur. 1999. *The law of the lifegivers*. Amsterdam: Harwood.

Dumon, D., and R. Devisch. 1991. *The oracle of Maama Tseembu*. Film, Belgian-Flemish Television: Science Division.

Ettinger, L.B. 2006. *The matrixial borderspace*. Minneapolis: University of Minnesota Press.

Henare, A., M. Holbraad, and S. Wastell, ed. 2007. *Thinking through things*. London: Routledge.

Langer, A., and A. Lutz, ed. 1999. *Orakel: Der Blick in die Zukunft*. Zürich: Museum Rietberg.

Merleau-Ponty, M. 1964. *Le visible et l'invisible*. Paris: Gallimard.

Van Binsbergen, W. 2003. *Intercultural encounters*. Berlin: LIT Verlag.

Van Binsbergen, W., ed. 2011. *Black Athena comes of age*. Berlin: LIT Verlag.

Viveiros de Castro, E. 2004. Exchanging perspectives: The transformation of objects into subjects in Amerindian ontologies. *Common Knowledge* 10, no. 3: 463–84.

Health care decisions by Sukuma 'peasant intellectuals': a case of radical empiricism?

Koen Stroeken

African Cultures, Ghent University, Belgium

Health care decisions in Sukuma-speaking rural communities in Tanzania reproduce a practical epistemology that could be described as radically empiricist, rather than just pluralist; their point of reference is the deeper 'relation' between events, which collective traditions articulate and subjects may experience, but which escapes the atomistic perception privileged by biomedicine. This analysis relies on a diverse portfolio of ethnographic data, including the use and structure of medicinal recipes, the choices of mental health care according to experienced 'effectiveness', and lay discussions on the correct aetiology and treatment of reproductive disorder. Combining two dimensions for a given medical epistemology, the (empirical/ habitual) basis of its transmission and the (open/closed) relation with other epistemologies, four types are proposed: monism, dualism, pluralism, and radical empiricism. The concept of peasant intellectuals, it is argued, needs to be rethought in contexts of medicinal initiation.

Feierman (1990, 18) was inspired by Gramsci when he used the term 'peasant intellectuals' to acknowledge the creative intellectual activity and the actual organization of political movement by a select group of rural inhabitants in Tanzania, who at the brink of Independence shaped the country's future by making use of their social position at the nexus between (colonial) domination and (rural) public discourse. This term is adopted here, yet with an important re-orientation to the life-world of Sukuma farmers, the largest peasant group of Tanzania (which Feierman briefly mentioned in his book). Peasant intellectuals are not a select group in Sukuma communities, because the system of decision-making is polycentric and consensual: rather than the government, the village or a majority of individuals, all extended families usually decide together, with each adult representing the family (*kaya*). In Sukuma postcolonial history, 'power' and 'intellectual' refer less to political influence than to medicinal and divinatory knowledge acquired by joining one or more initiatory systems, including new traditions developed by upcoming healers.

The widespread presence of such medicinally aware peasants in Tanzania can explain a recent counterintuitive finding by Mshana et al. (2008, 35). They observed that in urban Dar Es Salaam an affliction such as stroke 'was widely believed to emanate from supernatural causes (demons and witchcraft), while in rural Hai,

explanations drew mostly on "natural" causes (hypertension, fatty foods, stress).'
Contrary to conventional wisdom, the farmers thus seem to entertain an empirically
more sound position. The reason, as concluded from the author's fieldwork in
Tanzanian villages, is that farmers have more experience with the production of
medicine, in both making diagnoses and administering plant recipes. From
adolescence, men and women learn to put the effectiveness of 'magic' into
perspective. For example, although Sukuma-speaking villages are today witnessing
the dwindling of traditions, many adults have in their youth undergone a medicinal
initiation called *ihane*, which involved training sessions with plants in the fields and
which thus ensured an empirical approach to remedies. Their opinions about the
effectiveness of certain medicines derive from sensory experience rather than from
the dogma of 'traditional beliefs' or from the latter's counterpart, the science of
school curricula, which educated town-people must rely on. The rhythms of life and
work among urban dwellers are oriented around school, factory, sales, street or
company so most of them lack the farmers' daily experience of dealing with the
natural fertility of the land, crops and forest plants. Church sermons about the magic
foolishly embraced by so-called 'backward' villagers reinforce the idea of traditional
medicine belonging to an illicit realm, separate from daily life. In this paper,
parallelism or dualism (see Table 2, later) denotes such a closed relation to other
medical epistemologies and such belief-oriented transmission of a medical
epistemology. In town, one finds doctrines such as Christianity dominating public
discourse, with members proving their religious conversion in a sort of dualism that
leaves them fascinated with 'the occult' while fearing it. In the absence of initiation
rituals and daily medicinal routines, townspeople have less chance of developing a
cosmology based on practical experiences that could inform the advantages versus
disadvantages of traditional and biomedical therapies for certain illnesses.

That is how Mshana et al.'s surprising result can be explained. In the classic terms
of development anthropology (Hobart 1993), the newly educated in town are
structurally in a position of ignorance, having to rely on laboratories and accredited
experts for commonly available natural resources such as agricultural seeds or, in
this case, plant medicine. So, rather than practising opposite kinds of medicine in a
dual world, the rural and urban respondents score differently on a continuum
measuring the degree to which people are culturally enabled to develop their own
epistemology of healing and adapt it to personal experience.

The empirical in medical epistemologies: mental health care files

The empirical dimension of Sukuma rural epistemologies of healing becomes most
evident in their ongoing reliance on healers for those afflictions that Western
biomedicine is least able to treat, notably 'mental' illness. The affinity between
mental health and traditional healing is a general trend in Africa, as can be
established from a quick literature review.[1] In Tanzania's metropolis, Dar Es
Salaam, 48% of the 176 clients in Traditional Health Centres (an urban version of
the village healer's compound) suffered from mental illness, while among clients of
primary health care centres only half of that proportion did (Ngoma, Prince, and
Mann 2003). The first port of call for African patients suffering from mental illness is
often the traditional healer (cf. Yao et al. 2008, on Ivory Coast). Across rural Africa
a plurality of mental health care options exists, with faith consultations, traditional

healing and hospital services paralleling each other (cf. Adewuya and Makanjuola 2009, on Nigeria). Usually, mental illness is associated with a complex causality that may elude biomedicine and find expression in socio-experiential concepts such as 'spirits' and 'witchcraft' (cf. Igreja et al. 2010, on Mozambique; Shibre et al. 2008, on Ethiopia). This strong association between traditional healing and ailments eluding biomedicine should not be underestimated in the field of medical anthropology, long dominated by the concept of 'medical pluralism' (Slikkerveer 1982; Johannessen and Lázár 2006), because it enables us to attribute the fact of pluralism with the empirical basis of health care decisions. Since the affinity between mental health and healing rituals seems to hold for much of rural sub-Saharan Africa, medical anthropologists might consider rejecting pluralism as the default platform of rural health care decisions and develop new, more appropriate nuances within the empirical dimension.

Additional evidence comes from Uganda, where laypeople seem to make informed health care decisions *within* the list of mental illnesses. In a recent study by Abbo et al. (2009) among 387 patients of traditional healers, there was not only a bias towards treatment for mental illness in general (60.2%) but also towards consultations for symptoms that, according to the psychiatrist, pointed specifically to schizophrenia or other psychotic disorders (the prevalence rate of 29.7% was higher than expected). Since such disorders are particularly hard to treat in hospitals, with antipsychotics having many negative side effects (Bentall 2010), there may be an empirical foundation for clients trying out traditional healers with their wide arsenal of ritual therapies.

As the author's ethnographic research on traditional medicine in Sukuma villages progressed since 1995 to include more mixed methods such as quantitative data in 2010, indications have grown as to how empirically driven the health decision-making can be. In the mental health care files of Misungwi district hospital from 2009 (see Table 1), it appears that rural patients rarely consult for schizophrenia (11 out 147, or 7%) while they do for epilepsy and depression. A different picture emerges among urban patients in the same hospital. With a rate of 31% (15 out of 49), they frequently consult for schizophrenia (versus epilepsy and depression). The rural/urban contrast is statistically highly significant ($p < 0.0001$).[2] Might the contrast not reflect a lesser belief among villagers in hospital remedies for schizophrenia or what is called *mayabu*?

The villagers do go to the hospital for epilepsy (*lusalo*). So, far from avoiding the hospital or blindly trusting the healer, they seem to follow a particular epistemology, which discriminates between the afflictions of *lusalo* and *mayabu*. When asked, many know about anticonvulsants for epilepsy. Why then, in the case of schizophrenia, would they not go to the hospital for the antipsychotics? In-depth interviews among

Table 1. Three types of mental health diagnoses in Misungwi hospital (2009).

	Villages	Town	Total
Epilepsy	88	20	108
Schizophrenia	11	15	26
Depression	48	14	62

villagers confirm that, following their experiences with the successes and failures of both healers and hospitals, traditional healing is rumoured to be more effective in treating mental illness – especially *mayabu*, allegedly caused by witchcraft and manifesting itself in psychotic symptoms such as uncontrolled shouting, catatonia and aggressive behaviour. The association is strongest among Sukuma peasant intellectuals who have accomplished the *ihane* medicinal initiation and were registered in the hospital files with a *pagani* or pagan affiliation, also known as 'traditional religious affiliation' (a residual label reserved for patients who not only believe in ancestral spirits and magical protection but also would not associate with Islam, Catholic, Protestant or other denominations). Only one out of 20 *pagani* went to the hospital for what the records label 'schizophrenia'.

Radical empiricism in medicine: *shingila*

Mayabu is invariably claimed to have increased over the years with the introduction of mining and connection to the city after infrastructural works. *Nonga* medicine is said to 'restore the wits' (*kubeja masala*) of *mayabu* patients. Patients with the same symptoms live together at the healer's compound and twice daily receive two sorts of medicine, one of which is meant to counter the witchcraft. The other, *nonga* or 'the shell', is administered from a snail's shell into the patient's nose. The taste is so astringent that some vomit. It contains, among others, pungent pepper residues, which have fermented out in the sun in a soda-bottle hanging on the ancestral altars. Another ingredient is a plant called *fifi* (Artemisia afra Jacq.), commonly used against coughing, possibly in this case to reduce throat inflammation.

The main ingredient of *nonga* is the root of a caper plant (*busisi*), Capparis fascicularis (DC. *var* elaeagnoides). According to ethnopharmacological studies the plant has a significant immuno-stimulant effect of calming the brain's dopamine system, which may reduce psychotic symptoms (Rivera et al. 2003). Moreover, in a process known to botanists as the 'mustard oil bomb', caper plants contain enzymes that, when damaged by a herbivore, break down their glucosinolates into thiocyanates and isothiocyanates, among others. The first component has the potential, after contact with water, to remove cyanide from the body; the second component has been shown to inhibit the development of cancer and tumours (Grubb and Abel 2006).[3]

But the story does not end here. For Sukuma, healers and patients who completed their *ihane* initiation this is actually where the story begins, because for them the empirical (i.e., claims based on experience rather than on logic or belief) implies that medicine should contain, besides plants, an additive called *shingila*, literally 'entrance', which signals the medicine's purpose. In *nonga*, one *shingila* is a piece of a paled broom (*ikumbo lyape*), which expresses the objective of cleaning the body polluted with witchcraft. Another *shingila* is sheep urine, which may be associated with docile behaviour, not irrelevant for *mayabu*. The *ihane* initiation, which all Sukuma men are invited to undertake and which has a female equivalent (although both are disappearing under the pressures from churches and government), is basically a training into *shingila*: namely, the preparation of magical ingredients that give access ('entrance') to recovery by capturing the relation between illness, purpose and the larger world.

Defending a phenomenological approach to bring out this 'enskilment' in medicinal use as opposed to a dualist Cartesian epistemology, Hsu (2010: 28) posits: 'The idea that the self is intrinsic to and inseparable from perception, which phenomenology emphasizes, goes diametrically against the empiricist paradigm of perception.' Indeed, sensory evocative *shingila* are not empirical in the atomistic sense of Humean empiricism, which deals with things such as a disease, a body or a family, but in the *radically* empiricist sense, which considers the relations between these. Although these relations are invisible, people experience them as integrally part of their self.

For a radical empiricist such as William James (1975, 6–7), 'the relations between things, conjunctive as well as disjunctive, are just as much matters of direct particular experience, neither more so nor less so, than the things themselves. [. . .] the parts of experience hold together from next to next by relations that are themselves part of experience.' James hereby objects to 'the rooted rationalist belief that experience as immediately given is all disjunction and no conjunction' (James 1975, 6–7), which presupposes a knower separate from the things, using reason to acquire a truth that transcends temporal experience. For radical empiricists such (individual) reason comes after the event (or is a new event) and rather renders the original experience impure. The sense data obtained through our various modes of perception do not correspond to entities but are real through the relations experienced between the things. It is in the latter (experience) rather than the former (entities) that 'spirits' play an important role in traditional healing.

However, a phenomenology privileging subjectivity seems to have at least one blind spot: how can our observations of relations, of a self inseparable from its object, be put into words other than into those provided by our collective heritage of language? The *ihane* initiation by peasant intellectuals can be said to acknowledge that heritage by teaching the metaphorical language of *shingila*. For large multigenerational families who continuously deal with cases of affliction at home, the main point of reference for diagnosis and remedy is not the individual patient but the collective seeking to live with the fact of illness. Initiatory and other collective traditions possess a knowledge that no individual does. They evolve with people's changed experience of the meaningful relations between things, as illustrated next.

Medicinal intellectuals discussing diagnosis

Monitoring one's symptoms in relation to the effectiveness of a certain treatment, with the option of seeking help elsewhere, is central to Sukuma healing. Both clients and healers are used to a certain degree of failure, to times when the ancestor's wrath was too deep or the witch's schemes too clever to counteract. Much of the interaction between patients in the healer's compound concerns the proper diagnosis of their illness. That this remains an open question, a topic of ongoing conversation, should not be surprising since it is the healers themselves who from the start clarify to incoming patients that they rely on ancestral guides (*masamva*) and mediumship (*bumanga*) for diagnosis. They are bound by what the spirits inspire. Thus, healing is an open search, with not only healers and diviners but also a wide array of community members participating.

One morning in November when the air had been hot and humid as the rains were about to fall, an eight-month-old girl accompanied by her mother was admitted to the healer's compound where the author worked. The baby died around noon the same day. One assistant-healer expressed his regret to us that his sister, who for several years had been in charge of divinations (he and other siblings said they themselves could not since they never were possessed by the spirits like she was), had not managed to mount the spirit that morning. Otherwise she might have come up with the proper remedy. As the group conversed quietly, to not disturb the deceased's family members mourning nearby, a neighbour joined in to speculate on the cause of death. Did the mother not lose her previous child as well? He concluded that the mother suffered from either *loya*, 'little hair', or *masinzo*, 'scissors'. The two conditions are alike in their affecting of infants and in their endurance until the mother removes the cause. (Neither children's disease implicates witches). *Loya* refers to a little black hair on the mother's back that, soon after contact with the infant carried on the back, will lethally inflict the baby. The problem returns until removal of the fatal hair. The cause of *masinzo* is located in the mother's vagina, where two scissor-like bits of tissue would grow towards each other until about eight to ten months after birth when the 'scissors' close off the womb and in this way end the infant's life. (For Sukuma, a child's fate is intimately connected with that of the mother until several years after birth, so these early deaths are considered miscarriages.) Because the child died after eight months, *masinzo* seemed to our neighbour the most probable diagnosis.

The wife of the first speaker, the healer's son, disagreed. She reminded everyone that the woman's two previous infants died aged 1.5 and 2 years, which is too late for either 'little hair' or 'scissors'. It must be a *njimu*, she continued, a bad ancestor, who obstructs her fertility before or after pregnancy. *Njimu* refers to an ancestor who controls female fertility. The ancestor is located on a woman's matrilateral side, *ku migongo*, literally on the side 'of the backs'. The back has a strong maternal connotation. Failing to observe traditional rules at marriage such as dedicating a sheep to one's new wife may arouse a *njimu* curse. Her husband agreed and added that performing a collective ritual in honour of that *njimu* ancestor (who would have to be identified through oracles) could prevent this lady from losing any further children. That, he said, will be the probable outcome of discussions following the burial.

Some months later he talked differently about the ritual when his three-year-old girl Nkwimba got ill with high fever and severe convulsions (*nzoka ya ntwe*, 'snake of the head', better known in Swahili as *degedege*). Two mediumistic divinations, one by a diviner in the village of Mapilinga and one by his sister, revealed the demands of an ancestor who 'died in the wild', meaning outside the compound, which is considered the most ominous death. These oracles were independently confirmed by a chicken divination performed by the village elder Ngwana Chonja. The chicken oracle showed a double white outgrowth on the bird's back known as 'the two gourds' (*shisabo ibili*) which points to the role of an ancestor (who favours gourds over tin pots for food) and more exactly evokes his demand of ritually sacrificing two sheep in his honour. To the author's surprise, the healer's son chose not to follow the diviner's advice. 'Nowadays we, Sukuma, avoid sacrificial rituals because these do not prevent evil', he explained to the author in the company of his siblings. In funerals there is nothing wrong about a placating ceremony 'doing what will be refused anyway' (*kwita agalemagwa*). But in illness one has to be careful with interventions: 'The ancestors can get satiated. After the sacrifice they can lift their protection altogether and the child dies.' This tendency to immediate gratification instead of long-term exchanges of gifts was the general trend among Sukuma today, he further explained: in the past, the brideprice for marriage was paid in instalments, whereas nowadays the bride's family demands from the groom to 'complete once and for all' (*kumala gete*), that is, to hand over all the cattle during the wedding. He intuited a collective change in which the Sukuma and their ancestors participate.

Traditional cosmological concepts such as *loya*, *masinzo* and *njimu* structure people's health care decisions, whereby personal medical histories are verified. The 'truth' (*ng'hana*) is discussed in relation to past observations and new evidence. The relationship between ancestors, satisfaction and sacrifice makes therapeutic choices meaningful for the participants. And this relationship can change, as in the new aversion to sacrifice. To express his radically empiricist intuition, the healer's son referred to the collective. He voiced his health-care decisions in terms of 'we Sukuma', echoing the many medical conversations he regularly has with other peasant intellectuals in the family, the neighbourhood and beyond. An etiology attributing a woman's infertility to ancestral demands is radically empiricist in expressing the intuited relationship between affliction, remedy and communal peace. At a collective level such etiology is not less 'empirical' than a biomedical explanation. The latter could hardly prompt the community into a better response to infertility.

So, reading the signs of a current trend, the healer's son dropped the idea of ritual sacrifice. His wife continued what she had been doing all along to improve her fertility and in this way keep her child healthy, namely to daily smear her body downwards from top to toes with a medicinal preparation named *bugota wa njimu* (*njimu* medicine). Therapy thus consisted of engaging with medicinal plants, tapping into their intrinsic force without having to communicate with the ancestral spirits and depend on their accord. From this increased autonomy thanks to medicinal plants, it is a small step to preferring pharmaceutics. There can be no doubt that despite the clinical architecture, personnel, uniform and separate location by which hospitals stand out (thus materializing and institutionalizing a certain dualism with other medical services on the local pluralist market), biomedicine has much to offer that Sukuma patients desire today and that suits those hesitant about complex ancestral rites. Biomedicine attracts not so much for representing the West's expertise as for fitting in with the local, now transformed epistemology of healing (see Geissler and Prince 2010, 171). That epistemology is not fixed. Like Luo mothers in Kenya (Geissler and Prince 2010, 169), Sukuma respondents conceive of their existential condition as a continuous search. Yet, rooted in vibrant medicinal traditions and driven by empirical assessments of past interventions, as illustrated by the conversation above, it is a confident rather than Sisyphean search.

Synthesis: four types of medical cosmology

A radically empiricist epistemology of healing differs from the atomistic empiricism privileged by positivism, in that it acknowledges the 'relations' between things as part of human experience. Because these relations are less easily articulated, this epistemology of healing cannot be dogmatic, or dismissive of other medical epistemologies. The health-care decisions of Sukuma peasant intellectuals are inconclusive and inclusive, not because the empirical burden that they place on themselves is low, but rather to the contrary: for them the temporary calming effect of *nonga* (possibly due to the *busisi* plant's dopamine balancing) issues from a *shingila* that restores peace (*mhola*) at the more durable, socio-cosmological level. So, while staying in the hospital, patients will continue using such magical ingredients, clandestinely if necessary (cf. Langwick 2008).

This open relation to other epistemologies contrasts with the empiricism of the medical profession echoed by Angell and Kassirer (1998, 841): 'There cannot be two kinds of medicine – conventional and alternative. There is only medicine that has been adequately tested and medicine that has not, medicine that works and medicine that may or may not work. Once a treatment has been tested rigorously, it no longer matters whether it was considered alternative at the outset.' While thus cogently rejecting the dualism opposing biomedicine and alternative medicine, the authors opt for a monist take on medicine. A medicine will remain labelled as alternative (or traditional) when its effectiveness has not been proven in a lab, which places alternative medicine always – structurally – on the losing side: if it works it will eventually be called conventional. This is monist for denying the blind spot of scientific methodology. Not everything affecting health fits in a lab.

Like any social domain, biomedicine is epistemologically heterogeneous. Yet, as Stengers (1995) has remarked about the obsession of the medical profession with charlatans, a universalism subtends the medical sciences. The integration of alternatives has been scientifically fruitful but also attests to a closed relation with other epistemologies. Another type is the medical dualism of Christians, proving their conversion by denouncing traditional medicine (or some denominations forbidding biomedicine). The subjugation of other epistemologies by medical dualists (and monists) contrasts with the de facto medical pluralism of most patients in the world, many of whom appreciate the segregation of hospital and healer's compound, which actually permits pluralism. Medical pluralism is what most of us engage in who consume without obeying dogma or developing new remedies. We transmit the epistemology through cultural habit rather than experiential validation.

Sukuma peasant intellectuals belong to none of the above three types because, ever since their medicinal initiation, they have learned to assess health care services and contribute to the design of additives and healing rituals. In comparison to monism, with its universalist claim and separation of the experts from lay people, they display a constructivist (versus substantivist or hierarchical) approach to other epistemologies.[4] The Western counterpart of this fourth type may be psychotherapists willing to learn from healers (cf. Devisch 1993: Maiello 2008).

Table 2 schematizes four types of medical epistemology (or cosmology, defined broadly as an ordering of the world). In the left column of the table, a closed relation

Table 2. Four types of relation between, and transmission of, medical epistemologies.

Transmission	Relation	
	Closed	Open
Habitual	Dualism	Pluralism
Empirical	Monism	Radical empiricism

between medical epistemologies imposes hierarchy between the epistemologies (monism) as in positivism, or imposes segregation (dualism) as in certain Christian denominations. The spontaneous, de facto pluralism (above right) can be found in culturally diverse communities across the world, and also among clients with a post-modern epistemology. This pluralism will turn into radical empiricism, a fourth type, when the open approach to cosmologies is accompanied by decisions that are empirically based and mediated by the collective.

The author believes this fourfold and two-dimensional division of health care decisions to be more useful than current non-differentiations that implicitly condone dichotomies opposing biomedicine to traditional medicine, urban to rural users, Westerners to Africans, or universalists to pluralists.

Acknowledgements

The author is deeply indebted to the Sukuma traditional practitioners, especially Lukundula, and to Bon Peter, the mental health coordinator of Misungwi district, Tanzania. He thanks the Flemish Fund for Scientific Research and the BOF-fund of Ghent University for financing and ethically accrediting research in 2007 and 2010. The paper benefited much from the input of Elisabeth Hsu, Murray Last and Sushrut Jadhav. It was presented at the conference 'Medical Anthropology in Europe' funded by the Wellcome Trust and Royal Anthropological Institute.

Conflict of interest: none.

Notes

1. When searching on the Pubmed database with the terms 'Traditional medicine OR traditional healing OR traditional healers AND Africa AND mental' 205 papers in total were obtained. Forty-eight dated from 2000 until 2010. After exclusion on the basis of irrelevance (e.g., other afflictions such as stroke; 'mental' as a kind of effect; pharmacological study; other region than Africa), 31 papers were left, the main outcomes of which are presented briefly here.
2. The Chi-square is 17.4 with two degrees of freedom. The possibly higher prevalence of schizophrenia in urban areas cannot fully account for the urban-rural contrast (cf. Mortensen et al. 1999, measuring a disparity of schizophrenia risk with factor 2,4 between Denmark's largest city and the country's most rural area).
3. The author is indebted to Suzy Huysmans of the KU Leuven Laboratory of Plant Systematics for references and plant identification in 2006.
4. This does not refer to Horton's (1967) open predicament of 'Western modern science' seeking verification versus the closed predicament of 'African traditional thought' fearing falsification (an opposition of predicaments which the author's stance would then seem to invert; see also Trawick 1987). What mattered in this paper is the practical relation – observed in the practice of health seekers – to other epistemologies, and not the position taken by a belief or a theory in relation to data conflicting with it.

References

Abbo, C., S. Ekblad, P. Waako, E. Okello, and S. Musisi. 2009. The prevalence and severity of mental illnesses handled by traditional healers in two districts in Uganda. *African Health Sciences* 9: S16–22.
Adewuya, A., and R. Makanjuola. 2009. Preferred treatment for mental illness among Southwestern Nigerians. *Psychiatric Services* 60, no. 1: 121–4.

Angell, M., and J. Kassirer. 1998. Alternative medicine – the risks of untested and unregulated remedies. *New England Journal of Medicine* 339: 839–41.

Bentall, R. 2010. *Doctoring the mind: Why psychiatric treatment fail.* London: Penguin.

Devisch, R. 1993. *Weaving the threads of life: The Khita gyn-eco-logical healing cult among the Yaka.* Chicago: Chicago University Press.

Feierman, S. 1990. *Peasant intellectuals: Anthropology and history in Tanzania.* Madison: University of Wisconsin Press.

Geissler, P., and R. Prince. 2010. *The land is dying: Contingency, creativity and conflict in Western Kenya.* Oxford: Berghahn.

Grubb, C.D., and S. Abel. 2006. Glucosinolate metabolism and its control. *Trends in Plant Science* 11, no. 2: 89–100.

Hobart, M. 1993. *An anthropological critique of development: The growth of ignorance.* London: Routledge.

Horton, R. 1967. African traditional thought and Western science. *Africa* 37: 50–71.

Hsu, E. 2010. Introduction: Plants in medical practice and common sense. In *Plants, health, and healing: On the interface of ethnobotany and medical anthropology*, ed. E. Hsu and S. Harris, Oxford: Berghahn.

Igreja, V., B. Dias-Lambranca, D. Hershey, L. Racin, A. Richters, and R. Reis. 2010. The epidemiology of spirit possession in the aftermath of mass political violence in Mozambique. *Social Sciences and Medicine* 71, no. 3: 592–9.

James, W. 1975. *The meaning of truth.* Cambridge: Harvard University Press.

Johannessen, H., and I. Lázár, eds. 2006. *Multiple medical realities: Patients and healers in biomedical, alternative and traditional medicine.* Oxford: Berghahn.

Langwick, S. 2008. Articulate(d) bodies: Traditional medicine in a Tanzanian hospital. *American Ethnologist* 35, no. 3: 357–371.

Maiello, S. 2008. Encounter with a traditional healer: Western and African therapeutic approaches in dialogue. *Journal of Analytical Psychology* 53, no. 2: 241–60.

Mortensen, P., C. Pedersen, T. Westergaard, J. Wohlfahrt, H. Ewald, O. Mors, P. Andersen, and M. Melbye. 1999. Effects of family history and place and season of birth on the risk of schizophrenia. *The New England Journal of Medicine* 340, no. 8: 603–8.

Mshana, G., K. Hampshire, C. Panter-Brick, R. Walker, and The Tanzanian Stroke Incidence Project Team. 2008. Urban–rural contrasts in explanatory models and treatment-seeking behaviours for stroke in Tanzania. *Journal of Biosocial Science* 40: 35–52.

Ngoma, M., M. Prince, and A. Mann. 2003. Common mental disorders among those attending primary health clinics and traditional healers in urban Tanzania. *British Journal of Psychiatry* 183: 349–55.

Rivera, D., C. Inoncencio, C. Obon, and F. Alcaraz. 2003. Review of food and medicinal uses of capparis l. subgenus capparis (capparidaceae). *Economic Botany* 57, no. 4: 515–34.

Shibre, T., A. Spångéus, L. Henriksson, A. Negash, and L. Jacobsson. 2008. Traditional treatment of mental disorders in rural Ethiopia. *Ethiopian Medical Journal* 46, no. 1: 87–91.

Slikkerveer, L. 1982. Rural health development in Ethiopia: Problems of utilization of traditional healers. *Social Science & Medicine* 14, no. 21: 1859–72.

Stengers, I. 1995. Le médecin et le charlatan. In *Médecins et Sorciers: Les empêcheurs de penser en rond*, ed. T. Nathan and I. Stengers, Paris: Synthélabo.

Stroeken, K. 2010. *Moral power: The magic of witchcraft.* Oxford: Berghahn.

Trawick, M. 1987. The Ayurvedic physician as scientist. *Social Science & Medicine* 24, no. 12: 1031–50.

Yao, Y., Y. Yeo-Tenena, E. Tetchi, C. Assi-Sedji, Y. Bombo, L. Kouame, S. Soro, and R. Delafosse. 2008. First therapeutic recourse of the teenagers received to the service of Mental Hygiene of the INSP of Abidjan. *Mali Médicine* 23, no. 3: 55–60.

Index

Note: Page numbers in *italic* type refer to *tables*
Page numbers followed by 'n' refer to notes